T0331753

Nucleation and Growth of Metals

To my father

Series Editor
Robert Baptist

Nucleation and Growth of Metals

From Thin Films to Nanoparticles

Paul-Henri Haumesser

ELSEVIER

First published 2016 in Great Britain and the United States by ISTE Press Ltd and Elsevier Ltd

ISTE Press Ltd
27-37 St George's Road
London SW19 4EU
UK

www.iste.co.uk

Elsevier Ltd
The Boulevard, Langford Lane
Kidlington, Oxford, OX5 1GB
UK

www.elsevier.com

For information on all our publications visit our website at http://store.elsevier.com/

British Library Cataloguing-in-Publication Data
A CIP record for this book is available from the British Library
Library of Congress Cataloging in Publication Data
A catalog record for this book is available from the Library of Congress
ISBN 978-1-78548-092-8

Printed and bound in the UK and US

Contents

Acknowledgments ix

Introduction xi

Chapter 1. The Fabrication of Micro- and Nanostructures . . . 1

1.1. The fabrication of advanced interconnect structures in
microelectronic devices 1
  1.1.1. Ultralarge-scale integration 2
  1.1.2. The damascene architecture 3
  1.1.3. 3D integration 7
1.2. Elaboration of metallic NPs 8
1.3. Conclusions 12

Chapter 2. Phase Transition: Nucleation, Growth,
Aggregation and Coalescence 13

2.1. What is the easiest path for a phase transition? 13
  2.1.1. Gibbs energy of a homogeneous phase 13
  2.1.2. The driving force: supersaturation 14
  2.1.3. The resistance force: the energetic barrier 15
  2.1.4. Nucleus, subnucleus and supernucleus 16
  2.1.5. Phase transition: a step-by-step transformation 18
2.2. Nucleation 18
  2.2.1. General equations for nucleation 19
  2.2.2. Equilibrium (under non-equilibrium conditions) 21
  2.2.3. Nucleation rate 25
  2.2.4. Stationary nucleation 26

2.2.5. Classical description of clusters: the capillarity
approximation 27
2.2.6. Time-dependent nucleation 39
2.2.7. Limitations of the classical theory of nucleation 42
2.3. Growth 43
2.3.1. Attachment of monomers 44
2.3.2. Crystal growth 45
2.4. Aggregation 47
2.4.1. General equations for phase transformation 48
2.4.2. Aggregative growth 50
2.5. Coalescence 51
2.5.1. The Johnson–Mehl–Avrami–Kolmogorov equation 51
2.5.2. Exclusion zones 55
2.6. Conclusions 57

**Chapter 3. The Precipitation of Metals: Thin Film
Electroplating and Nanoparticle Synthesis** 59

3.1. Principles of metal electroplating 59
3.1.1. Thermodynamic considerations 60
3.1.2. Kinetics of electroplating and Faraday's Law 65
3.1.3. Electrochemical response of nucleation phenomena 66
3.2. Chemical synthesis of metallic nanoparticles 67
3.2.1. Mechanism of formation: the Lamer model 67
3.2.2. Mechanisms of stabilization 69
3.3. Conclusions 70

**Chapter 4. Copper Electroplating: from Superconformal
to Extreme Fill** 71

4.1. Copper electroplating 71
4.1.1. Specificities of Cu electroplating 71
4.1.2. Is electroplating capable of filling cavities? 72
4.2. Superconformal fill of damascene trenches 73
4.2.1. The electrolytes for superconformal Cu electroplating . . . 73
4.2.2. Suppressor and accelerator: an electrochemical study 74
4.2.3. The mechanism of superconformal fill 83
4.3. Extreme fill of TSVs 92
4.3.1. Is it easier to fill large TSVs than small damascene
structures? 93
4.3.2. Mass transport within TSVs and depletion of Cu^{2+} ions . . . 93
4.3.3. Electrolytes for TSV fill 96

4.3.4. Electrochemical study of an electrolyte capable of extreme
fill 99
4.3.5. The mechanism of extreme fill 102
4.4. Conclusions 105

Chapter 5. Nucleation and Growth of Metallic Thin Films . . . 107

5.1. Seed layer enhancement 107
5.1.1. Electroplating copper seed layers: which electrolyte? 107
5.1.2. First implementation of the SLE process in
shallow TSVs 110
5.1.3. Nucleation of seed layer enhancement (SLE) on resistive
barrier materials 111
5.1.4. Enhancement of nucleation density on resistive barriers . . . 112
5.1.5. Integration of the SLE process 116
5.2. Electroless deposition of self-aligned metallic barriers 119
5.2.1. Why would aligned metallic barriers be needed? 119
5.2.2. Principles of electroless deposition 121
5.2.3. Experimental conditions for electroless deposition 125
5.2.4. Deposition kinetics and film composition 127
5.2.5. Local thickness variations 129
5.2.6. Factors influencing the nucleation of CoWP and CoWB . . 130
5.2.7. Undesired, parasitic nucleation of CoWP and CoWB 134
5.2.8. Improvement of EM resistance 138
5.3. Conclusions 142

**Chapter 6. Nucleation and Stabilization of Metallic
Nanoparticles in Ionic Liquids** 143

6.1. What are ionic liquids? 143
6.2. Chemical synthesis of metallic NPs in ILs 146
6.2.1. Synthesis of Cu-NPs 148
6.2.2. Synthesis of Ta-NPs 149
6.3. Interactions between ILs and NPs 151
6.3.1. Viscous stabilization 151
6.3.2. Electrostatic interactions 152
6.3.3. Adsorption of ions 153
6.3.4. Impact of these interactions on nucleation, growth
and aggregation 155

6.4. Size evolution of Ru-NPs 156
6.4.1. Conventional size control 156
6.4.2. Unconventional size control 160
6.5. Conclusions 163

Appendices 165

**Appendix 1. Demonstration of the General Expression for
the Nucleation Rate** 167

**Appendix 2. Stationary Cluster Population and Nucleation
Rate** 169

Appendix 3. Where is the Surface of the Cluster? 173

Glossary 175

List of Acronyms 179

Bibliography 183

Index 193

Acknowledgments

As a scientist in CEA-Leti, I have been fortunate enough to be confronted with a variety of stimulating industrial challenges and academic questions in the same field of expertise: the precipitation of metals from solutions, either by electrochemical or chemical processes. I have tried to put in this book the essence of what I have learnt in this field.

I have learnt a lot from my colleagues. Some of them contributed in this work: Maryline Cordeau, Céline Jayet, Muriel Chaupin, Ségolène Olivier, Anne Roule, Thierry Mourier, Sylvain Maîtrejean, Olivier Pollet and Xavier Avale have all be of great help in the development of the electroplating and electroless processes. I will not forget Catherine Santini, my colleague in CPE Lyon, who has introduced me to the fascinating world of ionic liquids and who continues bringing new ideas in the research on nanoparticle (NP) synthesis. I am also grateful to Vincent Jousseaume, Frédéric Gaillard, Eric Chaînet and Daniel Lincot for sharing with me their deep scientific knowledge.

I have learnt a lot from my students. I thank them all: Sandrine DaSilva, Tifenn Decorps, Karim Sidi Ali Cherif, Mariana Amuntencei, Julien Cuzzocrea, Philippe Arquillière, Inga Helgadottir and Walid Darwich. Without them, life would have been much less enjoyable and this book would not have been possible.

Finally, I would like to thank my family: my children, Oscar, Léonard and Mélisande, and my wife, Sandrine. I thank them for their patience and support as deeply as I love them.

Introduction

Mastering the art of metal work was a decisive moment for mankind several thousand years ago. Since the middle of the last century, another decisive technological breakthrough changed our world. Microcomputing entered our homes more than 30 years ago. We have been carrying microcomputers in our pockets since the early 2000s. They are now invading our everyday life in a variety of "connected" items. This spectacular evolution mirrors the tremendous progress in the miniaturization of integrated circuits (ICs). These devices pack an ever increasing number of components in an ever smaller volume. As a result, the circuitry needed to interconnect these components becomes finer and more complex as technology progresses. The metallic conductors have now sub-micronic sizes. At the same time, new objects and applications are emerging in which metal is used as nanoparticles (NPs). As a result, *we must now master the art of metalwork at the microscale and even at the nanoscale.*

Since ancient times, metal has been shaped into the desired object using a *top-down* approach. A raw piece of metal is cut, hammered or molded into the final object. To fabricate the very small metallic structures in ICs, it is preferable to use a *bottom-up* approach in which metal is *grown* to its final shape. This is usually done by *precipitating* the metal from a solution containing suitable chemical precursors. For this reason, the art of metalswork at small scales would be more of an alchemist's role rather than a blacksmith's.

The purpose of this book is to explain the ways in which this precipitation reaction can be controlled to produce the desired object.

The nature, the size and the shape of the latter is dictated by its exact use. Conduction lines in ICs are quite long (up to several centimeters) but very narrow (down to several tens of nanometers). Metallic NPs are metallic clusters below 10 nm in diameter. In Chapter 1, both types of metallic objects are presented. The progress of IC technology is briefly described. To sustain the pace of circuit miniaturization, Cu has been introduced as a conduction metal in ICs. The various processes involved in the fabrication of advanced Cu interconnect structures and the associated challenges are reviewed for conventional integration as well as more recent three-dimensional stacking strategies. In the last section of Chapter 1, the need for metallic NPs is formalized, and the difficulties associated with the elaboration and stabilization of these nano-objects are discussed.

The precipitation of metal is the core process in the bottom-up fabrication of small metallic objects. It is a special case of phase transformation. Because such phase transformations are of paramount practical and technological importance, they have been extensively studied. In Chapter 2, the general concepts associated with phase transformation are introduced. A very simple thermodynamic analysis of the problem shows that it generally starts with the nucleation of extremely small domains of the new phase in the old phase. Hence, controlling this nucleation is the most efficient strategy to optimize bottom-up processes, especially when extremely small sizes are targeted. For this reason, the so-called *classical nucleation theory* is discussed in detail in this chapter. Controlling the growth, aggregation and coalescence of these nuclei is also mandatory, as is illustrated by practical examples in the subsequent chapters. Therefore, a brief theoretical description of these steps of phase transformation is also proposed in Chapter 2.

In Chapter 3, these general concepts are developed further in the more specific case of chemical and electrochemical precipitations of metals. In particular, the electrochemical mechanisms associated with Cu electroplating are discussed. Similarly, important aspects of the elaboration and stabilization of metallic NPs in solution are briefly exposed. Hence, both Chapters 2 and 3 should be a nice introduction for the student or the scientist to questions of nucleation, growth, aggregation and coalescence, especially when chemical precipitation reactions are involved.

Chapters 4–6 are dedicated to the description and discussion of very specific processes for the fabrication of small metallic objects. In Chapter 4, the controlled growth of Cu in electroplating processes is shown to allow for the so-called superconformal fill of holes and trenches. This particular regime is used to fabricate the most aggressive interconnect structures. In Chapter 5, two examples of thin film deposition by (electro)chemical reactions are described. These are thin Cu seed layers and metallic barriers also needed in the fabrication of the interconnect structures. Finally, a very original and promising approach to fabricate metallic NPs is developed in Chapter 6. This very simple process uses a new class of solvents, the ionic liquids (ILs).

These examples have been selected based on their relevance regarding the technological challenges mentioned in Chapter 1. They are all aimed at overcoming current limitations of existing processes. As such, they provide up-to-date information for the technologist interested in these questions. Throughout these chapters, it is also shown how the concepts introduced in Chapters 2 and 3 can be (at least qualitatively) used to guide the optimization of these new processes. As such, they should be exemplary to researchers involved in similar developments.

1

The Fabrication of Micro- and Nanostructures

As Feynman first stated in his visionary lecture in 1959 [FEY 92], there is plenty of room at the bottom. Indeed, the source of most of our current technological progresses is our ability to manipulate and shape matter at a very small size, down to the nanometer scale. In this chapter, it is not our purpose to extensively cover the progresses in microfabrication or the emergence of nanoscience. Rather, we shall illustrate current challenges in both fields through selected examples: the fabrication of advanced interconnect structures in microelectronic devices and the elaboration of metallic nanoparticles (NPs).

1.1. The fabrication of advanced interconnect structures in microelectronic devices

One of the most prominent discoveries during the 20th Century was the transistor invented by Bardeen and Brattain at the Bell Laboratories in 1947 [BAR 48]. Less praised, but not less important, is the first demonstration in the mid-1950s of an integrated circuit on a silicon substrate [KIL 64]. Indeed, this was the starting point of a still on-going technological race toward more integrated, thus more complex, circuits. As they get smaller and "smarter", these devices are no longer restricted to desk computers, but have been introduced in a variety of mobile applications (laptops, phones, tablets, GPS systems, etc.), cars, televisions, etc. With the advent of the "Internet of Things", they are meant to bring brains (or at least communication skills) to a much vaster number of objects in our everyday life.

1.1.1. Ultralarge-scale integration

Since the mid-1950s, this evolution has demanded to integrate more and more transistors per chip [MOO 65]. To do so, the size of the transistors has progressively been reduced, as well as the size of all other surrounding structures such as interconnects. By the mid- 1960s, the industry moved from small- to medium-scale integration (less than 1,000 transistors per chip), then to large-scale integration (10^3 to 10^5 transistors) in the early 1970s. Since 1983, the so-called very large scale integration scheme has been adopted, in which the number of transistors exceeds 10^5. More recently, the term ultralarge-scale integration has appeared for circuits containing more than 10^6 transistors.

As a result, the density of transistors has increased exponentially over the years, as predicted since the early 1960s by Moore [MOO 65]. According to Moore's law, the density of transistors doubles every 18 months (Figure 1.1 (left)).

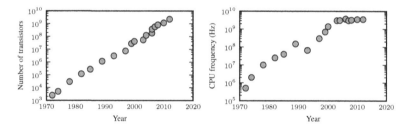

Figure 1.1. *Evolution of (left) the number of transistors per chip and (right) CPU clock frequency over the years*

Until the end of the 1990s, this trend was only sustained by the miniaturization of the transistors and interconnects, without any significant modification of their structure. In this period, the progress was measured by the ramp up of the device's clock frequency, which increased from $0.5\,\text{MHz}$ to $3\,\text{GHz}$ (Figure 1.1 (right)). Indeed, as the size of the transistors was reduced, so was their characteristic response time. However, in the meantime, the response time of the interconnect structures was increasing due to capacitive coupling. Soon enough, the delay time associated with the interconnects became limiting. Currently, the delay time associated with interconnect structures exceeds the response time of transistors by 3 orders of magnitude (Figure 1.2).

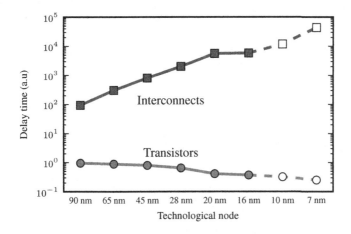

Figure 1.2. *Evolution of delays associated with transistors and interconnects as technology progresses. Adapted from [YEA 13]*

As shown in Figure 1.3, typical interconnect structures are formed by metal lines isolated by a dielectric material [LE 13]. Several levels of these lines are needed to form all the required interconnections in a circuit. Each of these "metal levels" is connected to the levels above and below by contact holes called *vias*.

Electrically, the lines are resistors (R). Adjacent, parallel metallic lines also form capacitors (C, see Figure 1.3). In its most simplistic description, such a circuit has a cut-off frequency given by:

$$f_{RC} = \frac{1}{2\pi RC} \qquad [1.1]$$

In other words, this circuit is expected not to be able to propagate signals whose frequency exceeds f_{RC}. To increase this cut-off frequency, it is thus necessary to either reduce R or C. In the late 1990s, chip manufacturers did both.

1.1.2. *The damascene architecture*

Until then, the metal and dielectric material used to fabricate interconnect structures were Al and SiO_2, respectively. To reduce R and C, these materials were abandoned and replaced by Cu and so-called *low-K* dielectric materials. The latter are SiO_2 derivatives, incorporating apolar chemical species such as methyl groups to decrease their relative permittivity [LE 13]. More recently, porous variants have been introduced: the incorporation of nanosized pores in the material allowed further reduction of the K value down to 2.2 [GRI 01].

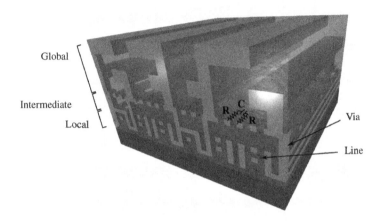

Figure 1.3. *Typical local arrangement of a microchip, including local, intermediate and global interconnects arranged in several "metal levels" connected to each other by vias*

1.1.2.1. Cu as a conducting metal

Cu was selected as a replacement metal for Al because of its lower resistivity ($\rho_{Cu} = 1.67$ $\mu\Omega\cdot$cm), but also because of its better resistance to electromigration (EM) as compared to the more conductive Ag and Au (see section 5.2.1 for more details about EM).

However, the introduction of Cu has not been straightforward. Indeed, this metal was not compatible with the process flow used to fabricate a metal level with Al. The latter is shown in Figure 1.4 (left). In a first step, Al is deposited on the substrate (which already has a metal level or contact plugs to the transistors if this is the first metal level). The connection structures (lines or vias) are then patterned into the metal. In this step, the excess metal is etched to leave only the final structures. Then, the dielectric material is deposited. A chemical mechanical polishing (CMP) process is finally applied to remove the excess insulating material and to planarize the metal level.

It turns out that there is no industrial process capable of properly etching Cu. For this reason, an alternative process flow was devised (Figure 1.4 (right)) [AND 98]. In this sequence, a layer of dielectric material is deposited first, in which the interconnect structures are etched. Then, the metal fills the trenches (lines) or holes (vias) and the excess metal is finally polished away. This architecture has been called *damascene*, after an ancient technique used in jewellery in which gold is interlaced into iron or steel [HES 07].

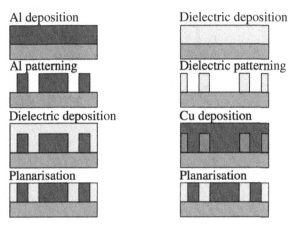

Figure 1.4. *Comparison between (left) the conventional integration of Al and (right) the damascene architecture*

The success of this approach, which was crucial to further improve the performances of microelectronic devices, depended on one condition: a process was required which was capable of filling the trenches and holes without leaving any void. This process was introduced by researchers at IBM in the late 1990s [AND 98]: it is Cu electroplating.

1.1.2.2. *Cu electroplating, seed layer and barrier deposition*

Conceptually, Cu electroplating is rather simple. It consists of electrolytically depositing copper onto the substrate from an electrolyte[1], essentially composed of Cu sulfate and sulfuric acid. However, it was found that by incorporating appropriate additives in this electrolyte a specific deposition regime takes place, in which deposition is significantly faster inside the features (Figure 1.5). This particular regime is referred to as

---

1 The principles of metal electroplating are discussed in section 3.1.

superconformal deposition [AND 98]. This superconformal regime is the only one capable of reliably filling the interconnect features and can be considered as the corner stone of the damascene approach. Its description is the main topic of section 4.2.

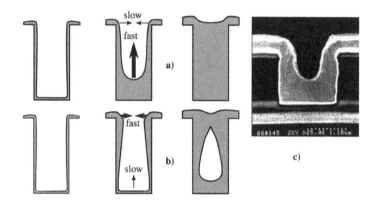

Figure 1.5. *Schematic representation of the a) superconformal; b) subconformal deposition regimes; c) early example of superconformal fill adapted from [AND 98]*

To initiate Cu electroplating, a conducting surface is needed. Therefore, a thin metallic layer is deposited first on the substrate. Usually, a Cu coating is used, deposited by physical vapor deposition (PVD) [NIS 07]. This thin Cu liner is called the *seed layer*. The elaboration of this thin metallic layer with an electroplating process alternative to PVD is discussed in section 5.1.

At this stage, the final difficulty remains with the use of Cu. Indeed, this metal is known to easily diffuse into other materials such as Si or its oxides. Also, Cu is a contaminant that creates deep levels in the electronic structure of Si, compromising the operation of transistors [IST 02]. In larger concentrations, Cu can even form compounds with Si and completely destroy the devices [NEW 82]. For these reasons, it is mandatory to confine Cu within the interconnect structures. This is why specific barrier layers have been added in the structures to prevent Cu diffusion. The first one is deposited before the metal (i.e. before the Cu seed layer), and is usually a bilayer of TaN and Ta. This barrier is deposited by PVD, usually in the same equipment as for the seed layer[2]. The second barrier is deposited once the metal level is formed, after the polishing step. This upper barrier is usually formed by a dielectric material (typically a Si nitride and/or carbide) deposited by plasma-enhanced chemical vapor deposition [LE 13]. Recently, a metallic coating has been proposed to advantageously replace this layer. This will be the subject of section 5.2.

Finally, the conventional metallization sequence comprising barrier and seed deposition, Cu electroplating, polishing and encapsulation is depicted in Figure 1.6.

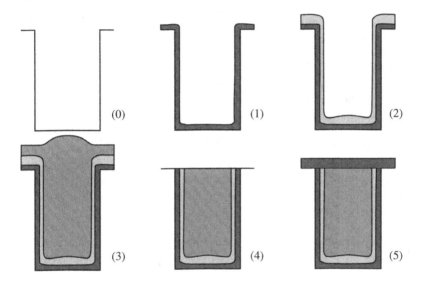

Figure 1.6. *Schematic representation of the metallization sequence of hollow trenches and vias including (1) barrier deposition, (2) seed layer deposition, (3) superconformal Cu fill, (4) CMP and (5) encapsulation*

1.1.3. *3D integration*

The co-integration of low-K dielectric materials and Cu is still used today in the industry. Because of this architectural modification, the device

---

2 This is needed to avoid air exposure of the barrier, which would result in the oxidation of the Ta-based barrier, see section 5.1.4.

manufacturers have been able to pursue Moore's law in recent years [LE 13]. However, as characteristic dimensions are shrinking down to a few (tens of) nanometers, physical limits are being approached.

For instance, the width of the Cu lines is reduced to 40 nm in the most recent technologies [ITR 13]. This dimension becomes comparable with the mean free path of electrons in Cu, which is about 50 nm at 293 K [HAN 02]. This means that electrons are no longer moving in an infinite medium; the probability that they collide with a wall becomes significant. This phenomenon causes an increase in the resistivity of the Cu lines. In such narrow lines, the resistivity of Cu can exceed $3\mu\Omega\cdot$cm [HAU 06].

Because of these limitations, pursuing Moore's law by miniaturization only is becoming more and more difficult and will even be impossible in the foreseeable future. For this reason, chip manufacturers are devising an alternative approach, referred to as three-dimensional 3D integration [LED 08]. This approach is schematically explained in Figure 1.7. The concept is quite simple: several layers of transistors (or any other device) can be stacked to increase their number per unit surface area. In addition, this significantly shortens the "long distance" interconnects in chips, resulting in improved performances. This strategy is conceptually simple and elegant, but raises technological challenges, such as drilling and metallizing deep contact holes to connect adjacent layers. These structures, called through silicon vias (TSVs), are key enablers for the 3D integration [KAT 10].

The metallization of TSVs proceeds through a similar sequence as for damascene interconnects (Figure 1.6). However, all the deposition processes, including barrier, seed layer and Cu electroplating need to be revised to comply with the depth of TSVs, as will be discussed in sections 4.3 and 5.1.

1.2. Elaboration of metallic NPs

Today nanosciences are a major field of research. Indeed, objects at the nanoscale (whose size ranges from 1 to 10 nm) are at the crossroads between the molecules manipulated by chemists and the technological devices fabricated by technologists (Figure 1.8). These different scientific fields are now converging in the exploration of this fascinating "nanoworld" [BAL 05].

Nano-objects possess properties between the bulk material and the molecule. For instance, metallic NPs offer optical properties that cannot be obtained from bulk metals. These properties have been used since antiquity, like for instance in the famous Lycurgus cup whose color changes from green to red depending on the incidence of light (Figure 1.9). This dichroic behavior is due to the presence of gold NPs in the glass.

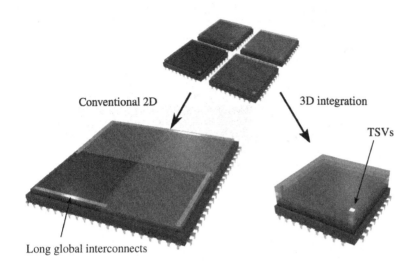

Figure 1.7. *Schematic representation of 3D integration as compared to the conventional 2D layout. The long interconnect lines in 2D conventional assembly are replaced by TSVs in the 3D integration. For a color version of this figure, see www.iste.co.uk/haumesser/metals.zip*

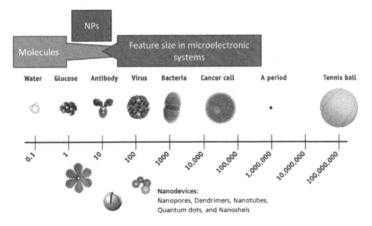

Figure 1.8. *The convergence of chemistry and technology in the "nanoworld"*

Figure 1.9. *The Lycurgus cup, fourth century – British Museum, London*

Moreover, these unique characteristics are tunable with the size and shape of these nano-objects. For instance, their color may vary as their diameter decreases (the same for their fluorescence, see Figure 1.10). This is because their physical properties evolve with their size, either by classical or quantum effects. In addition, NPs usually exhibit enhanced or unexpected chemical reactivity because they possess a significant amount of surface atoms. The fraction of surface atoms is called *dispersion* and scales with the inverse diameter of the NP [ROD 06]. Below 3 nm, surface atoms typically become the main constituents of NPs (Figure 1.11). Because surface and edge atoms have fewer neighbors as compared to the inner atoms, they are more prone to interact with other surrounding species. This ability may be used to enhance or modify the reactivity of the latter, which is the basis of the catalytic properties of metallic NPs [FRE 11]. It is also responsible for the chemisorption of compounds on metallic NPs, which is used to stabilize colloidal suspensions (see section 3.2.2).

Today, the versatile properties of metallic NPs are used in a variety of applications, such as medical imaging (using luminescent or magnetic properties), pigments, (bio)sensors, energy conversion and storage, and catalysis [GOE 10].

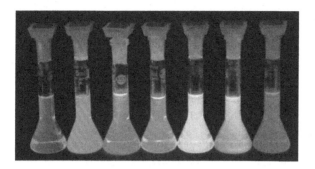

Figure 1.10. *Fluorescence of CdSe-CdS core shell NPs with a diameter of 1.7 nm (blue) up to 6 nm (red). For a color version of this figure, see www.iste.co.uk/haumesser/metals.zip*

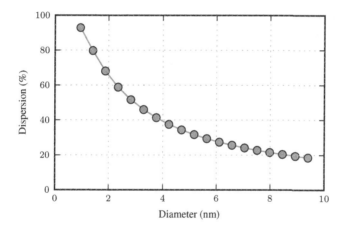

Figure 1.11. *Evolution of the dispersion with size for icosahedral Ru-NPs*

In all these applications, the size of the NPs must be accurately controlled to comply with the desired properties. This means that not only the average size needed for a given application is respected, but also that size distribution is as narrow as possible to ensure homogeneous performances.

There are two main approaches to elaborate NPs. In the first one, called *top-down*, the nanostructures are carved out of the bulk material. This may be achieved either by extensive grinding, or using more sophisticated approaches involving lithography patterning. The former is prone to contamination and agglomeration of the NPs. Besides, NPs below 50 nm are only marginally reachable. With the latter techniques, major drawbacks are their cost and the waste of removed material.

Hence, a preferred strategy is the *bottom-up* elaboration of NPs by combining atoms. This can be achieved either from the gas or the liquid phase at a moderate to low cost. In most cases, these processes yield quite pure NPs. However, the control of size and the tendency to agglomeration remain challenging. The elaboration and stabilization of metallic NPs from liquid phase processes is further described in section 3.2. A specific and innovative approach to this question is the object of Chapter 6.

1.3. Conclusions

The fabrication of metallic objects at the small size, either thin films or NPs, is of paramount importance for current and future technological applications. This elaboration is preferentially carried out using bottom-up approaches, in which the structures are assembled from individual atoms. Most of these processes must be controlled at the nanometer scale. This is mandatory either to obtain NPs with predictable size, or form thin films with optimized structure and properties.

As will be shown Chapter 2, the transformation of isolated atoms in a fluid phase[3] into bulk metal usually proceeds through the nucleation of small clusters that grow and coalesce into the dense metal. This is the case for the precipitation of a metal from a solution containing suitable precursors, which is the transformation involved in either the (electro)chemical deposition of metallic thin films or the chemical synthesis of metallic NPs. Fortunately, there exist many ways in which this nucleation–growth–coalescence sequence can be modified by tuning experimental conditions. The reasons why are explained in Chapters 2 and 3. In the subsequent chapters, these concepts will be applied to processes of technological interest in the fabrication of interconnect structures (Chapters 4 and 5) or to the synthesis of metallic NPs with accurate size control (Chapter 6).

---

3 The fluid phase may be a melted, but also a gaseous or liquid, solution.

Phase Transition: Nucleation, Growth, Aggregation and Coalescence

In this chapter, the general question of phase transformation is discussed from a quite fundamental standpoint. The notion of *nucleation* is introduced. This concept is central to understand and control the solidification processes described in Chapters 5 and 6 for the fabrication of thin films or nanoparticles. The subsequent steps of phase transformation, growth, aggregation and coalescence are also discussed. A large part of this discussion is inspired from the recent book by Kashchiev [KAS 00]. Here, only the essential concepts and equations are retained and updated whenever necessary.

2.1. What is the easiest path for a phase transition?

Phase transitions are profound transformations of matter. Like other natural processes, they must follow the easiest path. Therefore, it is a good idea to figure out what this path might be. In such a situation, thermodynamic considerations are usually extremely useful. In this section, basic thermodynamic notions will be recalled. Then, they will be used to identify the driving force for phase transition. Resistances against transformation will be considered as well. Finally, a general discussion of the whole process will be proposed.

2.1.1. *Gibbs energy of a homogeneous phase*

A phase α can be thermodynamically characterized by its entropy S_α, volume V_α and composition. The latter is described by the quantities $n_{\alpha,i}$ of the various chemical species that constitute the phase. Its energy U_α is related to these variables through [GUG 85]:

$$
\begin{aligned}
dU_\alpha &= TdS_\alpha - PdV_\alpha + \sum_i \left(\frac{\partial U_\alpha}{\partial n_{\alpha,i}} \right) dn_{\alpha,i} \\
&= TdS_\alpha - PdV_\alpha + \sum_i \mu_{\alpha,i} dn_{\alpha,i}
\end{aligned}
\qquad [2.1]
$$

where T and P are the temperature and pressure, respectively, and

$$
\mu_{\alpha,i} = \left(\frac{\partial U_\alpha}{\partial n_{\alpha,i}} \right)_{S_\alpha V_\alpha n_{j,\alpha}} \qquad (j \neq i) \qquad [2.2]
$$

is the *chemical potential* of species i in phase α. However, it is more practical to use the so-called *Gibbs energy* G_α as a characteristic function of phase α. This function is defined by:

$$
dG_\alpha = -S_\alpha dT + V_\alpha dP + \sum_i \mu_{\alpha,i} dn_{\alpha,i} \qquad [2.3]
$$

Since U_α is homogeneous of first degree, it is possible to use Euler's theorem to obtain:

$$
G_\alpha = U_\alpha - TS_\alpha + PV_\alpha = \sum_i n_{\alpha,i} \mu_{\alpha,i} \qquad [2.4]
$$

2.1.2. The driving force: supersaturation

Let us consider the transformation of a homogeneous phase into another homogeneous phase. Let us agree to call the initial phase "old" and the final phase "new". The transformation of the old phase into the new phase is possible only if:

$$
\Delta G = G_{new} - G_{old} < 0 \qquad [2.5]
$$

Let us limit ourselves to the case where this transformation concerns *a single species*, and let us agree to call the unit building block of the new phase (which can be an atom, a molecule, etc.) a *monomer*. If the transformation involves M monomers, the Gibbs energies can be related to their chemical potentials in the old and new phases:

$$\frac{\Delta G}{M} = \frac{G_{new} - G_{old}}{M} = \mu_{new} - \mu_{old} = \Delta\mu < 0 \qquad [2.6]$$

By definition, $\Delta\mu$ is called supersaturation. It corresponds to the favorable balance in Gibbs energy, which drives the phase transformation. The larger (the more negative) this difference, the more favorable will be the phase transformation.

2.1.3. The resistance force: the energetic barrier

Now comes the question of the actual path followed by the system to transform the old phase into the new phase. This transition necessarily proceeds by some continuous evolution of the system from the initial (old phase) to the final state (new phase). In other words, the system has to go through a succession of intermediate states. Generally, these states happen to be less stable than the old phase, so that $\mu_{inter} > \mu_{old}$ (or $G_{inter} > G_{old}$, see Figure 2.1). In other words, the old phase corresponds to a *local minimum of the Gibbs energy* (the true minimum corresponding to the new phase). For this reason, the old phase is said to be in a *metastable state*.

The system needs to overcome an energetic barrier in order to transform the old phase into the new phase, and this barrier plays a central role in the phase transformation.

From these simple considerations, an extremely important conclusion can be drawn: *the transformation is highly unlikely to involve the M species at once*. Indeed, the associated barrier would be $M(\mu_{inter} - \mu_{old})$. By comparison, for an alternative pathway involving a small number of species $n \ll M$, the associated energetic barrier is only $n(\mu_{inter} - \mu_{old})$. If n is really small as compared to M (several orders of magnitude in most practical cases), the latter barrier is by far easier to overcome. For this reason, first-order transformations usually proceed by *nucleation* (formation of small aggregates or *clusters* of the new phase), which subsequently *grow* to progressively transform the old phase into the new phase.

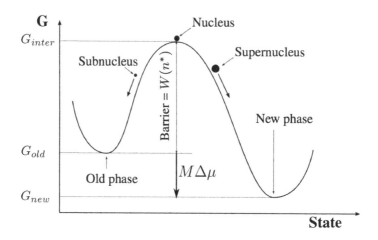

Figure 2.1. *Schematic representation of the transition path from the old to the new phase*

By definition, nucleation is the process by which extremely small (usually nanometric) aggregates (or clusters) of the new phase are formed, which irreversibly grow into macroscopic domains of the new phase. These clusters may form in the bulk of the old phase (homogeneous nucleation), but also at the interface with a substrate phase (heterogeneous nucleation).

Consequently, to properly describe the phase transition, it is necessary to determine the characteristics of these clusters. Again, this is equivalent to finding the *easiest* path (with smallest energetic barrier) between the old and the new phases. This is precisely the purpose of the various theories of nucleation.

2.1.4. *Nucleus, subnucleus and supernucleus*

Indeed, the granular theories of nucleation all aim at determining the work needed to form a cluster of a given size (expressed here by the number n of monomers contained in the cluster). This work corresponds to the difference in Gibbs energy between the initial state (homogeneous old phase) and the intermediate state (a cluster of n monomers embedded in the old phase, see Figure 2.1):

$$W(n) = G_{inter}(n) - G_{old} \qquad [2.7]$$

Remarkably enough, most of the nucleation theories lead to a general expression of this work as the balance between supersaturation and energetic penalties [KAS 00]:

$$W(n) = n\Delta\mu + \Phi(n) \qquad [2.8]$$

The first term corresponds to the gain in Gibbs energy associated with the formation of the n-sized cluster in the supersaturated old phase. The second term gathers all contributions resisting the nucleation (i.e. all phenomena associated with the formation of the cluster that cause an increase in the system's energy).

On the basis of this description, the energetic barrier to be overcome for the phase transformation is clearly the maximum value of this work. This value, noted W^*, corresponds to the formation of a cluster with a size n^* of the new phase. This specific size is called *critical*, and such a cluster is referred to as a *nucleus* (Figure 2.1). Indeed, any smaller cluster ($n < n^*$) will spontaneously redissolve (back to the local energy minimum, i.e. the old phase), whereas any larger cluster ($n > n^*$) will grow to form more of the new phase.

There is a critical size n^* (corresponding to the nucleus) below which clusters are non-viable and redissolve ($n < n^*$, subnuclei), and above which clusters ($n > n^*$, supernuclei) can grow (Figure 2.1). This critical size is determined by the condition of maximum $W(n)$:

$$\frac{dW(n)}{dn}\bigg|_{(n=n^*)} = 0 \qquad [2.9]$$

In this case, equation [2.8] becomes:

$$W^* = n^*\Delta\mu + \Phi(n^*) \qquad [2.10]$$

W^* is called *work of nucleation*.

2.1.5. *Phase transition: a step-by-step transformation*

From the considerations above, nucleation has been identified as the preferred process to transform the old phase into the new phase. However, nucleation is just a start. The nuclei are usually very small, and in many cases a lot of monomers remain in the old phase. Therefore, a significant portion of the transformation implies the growth of supernuclei. As their size and number increase, they start to coalesce, until all monomers are consumed. These different stages are schematically depicted in Figure 2.2. Our goal in the following sections is to study each of these processes as well as their interplay.

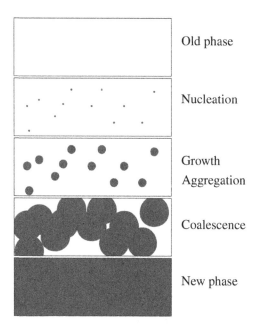

Figure 2.2. *The different stages of phase transformation*

2.2. Nucleation

The concept of nucleation has been introduced in the previous section on the basis of energetic considerations. Here, we shall describe this phenomenon in terms of elementary processes. We purposely restrict ourselves to the so-called *classical nucleation theory*, which is sufficient to account for the main phenomena relevant to the practical cases described in Chapters 4–6. The central idea in this model is that clusters (either subcritical, critical or supercritical) only grow (decay) by attachment (detachment) of monomers from (to) the old phase. The main result of this theory is the expression of nucleation rates in stationary and transient conditions. However, the practical evaluation of these nucleation rates requires that key parameters of the model, such as attachment and detachment frequencies or work for cluster formation, are known. In the classical theory of nucleation, the latter is evaluated using the *capillarity approximation*, in which the cluster is considered as a droplet of the new phase embedded in the old phase.

2.2.1. *General equations for nucleation*

During nucleation, the clusters of the new phase are very small and quite dilute. Therefore, they virtually do not interact, and *phenomena such as aggregation and coalescence are usually neglected*[1]. This implies that clusters mainly grow and decay by attachment/detachment of monomers only. Finally, the model for nucleation is based on a few simple hypotheses:

– the system is composed of clusters with variable size n ($n \geq 1$) embedded in the old phase (supported or not by a substrate, depending on the type of nucleation under consideration);

– for a given size n, all clusters have the same shape. Their concentration is $Z_n(t)$, and they are spatially homogeneously dispersed;

– *an n-sized cluster may grow by attachment of a monomer only*. This occurs at a frequency noted $F_n(t)$. Conversely, *the cluster may only decay by detachment of a monomer*. The associated frequency is noted $B_n(t)$. Both $F_n(t)$ and $B_n(t)$ may be time dependent;

– the system is not necessarily closed: n-sized clusters may additionally appear or disappear at rates noted $K_n(t)$ and $L_n(t)$, respectively.

In this model, clusters (either subcritical, critical or supercritical) only grow (decay) by attachment (detachment) of monomers from (to) the old phase. Moreover, in this description, monomers are considered as clusters with size $n = 1$.

---

1 Aggregation and coalescence are discussed in sections 2.4 and 2.5, respectively.

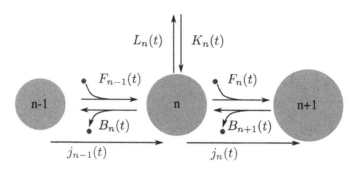

Figure 2.3. *The elementary processes of evolution for an n-sized cluster*

As a result, an n-sized cluster may only be produced by the growth of an $(n-1)$-sized cluster, decay of an $(n+1)$-sized cluster, or introduction in the system already formed at a rate $K_n(t)$. The same cluster may be eliminated from the system by the reverse processes (Figure 2.3). Finally, the evolution of cluster concentration is given by (for $1 < n < M - 1$):

$$\begin{aligned}
\frac{dZ_n(t)}{dt} &= F_{n-1}(t)Z_{n-1}(t) - B_n(t)Z_n(t) + B_{n+1}(t)Z_{n+1}(t) \\
&\quad - F_n(t)Z_n(t) + K_n(t) - L_n(t) \\
&= j_{n-1}(t) - j_n(t) + K_n(t) - L_n(t)
\end{aligned} \qquad [2.11]$$

where $j_n(t)$ represents the flux of clusters "crossing" the size n to larger sizes, given by:

$$j_n(t) = F_n(t)Z_n(t) - B_{n+1}(t)Z_{n+1}(t) \qquad [2.12]$$

To simplify the mathematical treatment of this problem, an elegant and efficient strategy consists of introducing the functions $C_n(t)$ that verify [KAS 00]:

$$F_n(t)C_n(t) = B_{n+1}(t)C_{n+1}(t) \qquad [2.13]$$

The nature of these functions will be discussed in the next section. For now, they help us to transform equation [2.11] into:

$$\frac{dZ_n(t)}{dt} = F_{n-1}(t)C_{n-1}(t)\left(\frac{Z_{n-1}(t)}{C_{n-1}(t)} - \frac{Z_n(t)}{C_n(t)}\right)$$

$$-F_n(t)C_n(t)\left(\frac{Z_n(t)}{C_n(t)} - \frac{Z_{n+1}(t)}{C_{n+1}(t)}\right) + K_n(t) - L_n(t)$$

[2.14]

Similarly, $j_n(t)$ can be expressed as:

$$j_n(t) = F_n(t)C_n(t)\left(\frac{Z_n(t)}{C_n(t)} - \frac{Z_{n+1}(t)}{C_{n+1}(t)}\right)$$

[2.15]

The concentrations of clusters as a function of their size and their evolution with time is described by a set of $M-1$ equations [2.11] or [2.14]. Equivalently, nucleation can be described by a set of $M-1$ equations [2.12] or [2.15] – expressing the flux of clusters "through" the various sizes.

2.2.2. Equilibrium (under non-equilibrium conditions)

The introduction of the functions $C_n(t)$ is an artificial procedure to transform the general equations for nucleation. However, these functions are not completely disconnected from the physics of the process, as we shall see now.

Let us consider the situation of a closed system ($K_n(t) = L_n(t) = 0$) in which the following condition is verified for all n values, at all times:

$$j_n(t) = 0$$

[2.16]

According to equation [2.12], this implies that:

$$F_n(t)Z_n(t) - B_{n+1}(t)Z_{n+1}(t) = 0$$

[2.17]

This means that for all sizes, at all times, the net result of monomer attachment and detachment is perfectly balanced. This corresponds to a situation of equilibrium[2]. Taking into account equation [2.13], the solution to this problem is obvious. At all times, we have:

$$Z_n(t) = C_n(t) \quad (1 \leq n \leq M) \tag{2.18}$$

As a direct result of equation [2.13], these concentrations can be usefully expressed as:

$$C_n(t) = \frac{F_1(t)F_2(t) \cdots F_{n-1}(t)}{B_{2(t)}B_3(t) \cdots B_n(t)} C_1(t) \tag{2.19}$$

A second result of equation [2.18] is that equation [2.14] readily yields (it is recalled that in the quasi-equilibrium situation, $K_n(t) = L_n(t) = 0$):

$$\frac{dZ_n(t)}{dt} = 0 \tag{2.20}$$

Therefore and as expected, in the quasi-equilibrium state, the cluster distribution does not vary over time.

Equation [2.18] shows that the $C_n(t)$ functions correspond to the **(quasi) equilibrium concentrations of the clusters. These functions have been introduced in the first place as a mathematical artifice to facilitate the treatment of the problem. In fact, even though we have been able to relate them to physical quantities,** *they remain artificial, simply because* **the system is not, under the conditions that prevail during nucleation, at equilibrium.**

Nevertheless, these concentrations have been used by many authors to derive a very useful relationship between attachment and detachment frequencies. Here, this development is introduced by considering each

---

2 More precisely, this describes a *quasi-equilibrium* situation, as the attachment and detachment frequencies are still allowed to vary with time. Under quasi-equilibrium conditions, the system instantaneously evolves to accommodate the variations of $F_n(t)$ and $B_n(t)$ to verify equation [2.17] at all times.

n-sized cluster as a chemical entity of its own, with its chemical potential μ_n. The thermodynamic equilibrium is reached when:

$$\mu_n = n\mu_{old} \tag{2.21}$$

To determine μ_n, we need to know $G(n)$, the Gibbs energy of a single cluster, which is given by:

$$G(n) = G_{inter}(n) - (M - n)\mu_{old} \tag{2.22}$$

With the help of equation [2.7], and recalling that $G_{old} = M\mu_{old}$, we find that:

$$G(n) = W(n) + n\mu_{old} \tag{2.23}$$

If all the clusters had the same size n, then $\mu_n = G(n)$. However, the system can contain clusters with various sizes, which again are considered as separate chemical species with their own chemical potential. Therefore, the chemical potential μ_n also contains an entropic mixing term [GUG 85]:

$$\mu_n = G(n) + kT\ln\left(\frac{C_n}{C_0}\right) \tag{2.24}$$

where C_0 is the concentration of sites available for cluster formation. This expression is only valid if the interactions between clusters are neglected (ideal mixing), which is reasonable considering their small size. Combining equations [2.23] and [2.24], we obtain:

$$\mu_n = W(n) + n\mu_{old} + kT\ln\left(\frac{C_n}{C_0}\right) \tag{2.25}$$

The equilibrium condition equation [2.21] thus leads to:

$$C_n = C_0\exp\left[-\frac{W(n)}{kT}\right] \tag{2.26}$$

which describes the concentration distribution of the clusters as a function of their size n. A useful alternative expression can be derived, in which C_n is related to C_1:

$$C_n = C_1 \exp\left[-\frac{W(n) - W(1)}{kT}\right] \qquad [2.27]$$

Injecting equation [2.8] in both equations yields the general expressions:

$$C_n = C_0 \exp\left[-\frac{\Phi(n)}{kT}\right] \exp\left[-\frac{n\Delta\mu}{kT}\right] \qquad [2.28]$$

$$C_n = C_1 \exp\left[-\frac{\Phi(n) - \Phi(1)}{kT}\right] \exp\left[-\frac{(n-1)\Delta\mu}{kT}\right] \qquad [2.29]$$

When applied to the undersaturated system, equations [2.28] and [2.29] accurately describe the distribution of cluster concentrations, because the system is at equilibrium. For the supersaturated system, the cluster population drops as n increases and reaches a minimum at $n = n^*$, as expected[3]. After that however, $C_n \to +\infty$ as $n \to +\infty$. Therefore, at least for large n values, the description is no longer physically correct, as already anticipated. Nevertheless, even in this case, these functions are the correct solutions to the "equilibrium" situation implied by equation [2.17], which results in the following relation between the detachment and attachment frequencies [KAS 00]:

$$B_n(t) = F_{n-1}(t) \exp\left(\frac{W(n, t) - W(n-1, t)}{kT}\right) \qquad [2.30]$$

This equation is general and does actually not depend on the particular form of $F_{n-1}(t)$[4]. The latter will be discussed in section 2.3.

---

3 In fact, the C_n functions are good approximates of the real cluster distribution in supersaturated systems for $n < n^*$ [KAS 00].

4 However, the grounds used here to derive this relation are unphysical, because the system is out of equilibrium during nucleation. Fortunately, it has been recently shown that a more rigorous treatment of this question leads to the same relation between attachment and detachment frequencies [SCH 05]. The interested reader will find all the details in the cited reference.

Equation [2.30] is a central simplification in the mathematical treatment of the problem, as it eliminates $M - 1$ parameters in the system of equations.

An interesting observation can be made if equation [2.30] is introduced in equation [2.19]:

$$W(n,t) = W(1,t) - kT(t) \sum_{m=1}^{n-1} \ln \left(\frac{F_m(t)}{B_{m+1}(t)} \right) \qquad [2.31]$$

This equation is an expression of the work of cluster formation entirely based on kinetic grounds: the work is expressed as a function of the attachment and detachment frequencies only. The evaluation of this work from energetic (or thermodynamic) considerations is discussed in section 2.2.5.6. For now, this relation can be used to predict the existence of the nucleus. Indeed, applying equation [2.9] to equation [2.31] yields:

$$F_{n^*}(t) = B_{n^*}(t) \qquad [2.32]$$

From a kinetic standpoint, the nucleus is this cluster for which the attachment and detachment frequencies of monomers are the same.

2.2.3. Nucleation rate

The nucleation rate $J(t)$ is defined in its most general way as the number of new supernuclei formed per unit time and unit volume (or surface):

$$J(t) = \frac{d}{dt} \left(\sum_{n^*(t)+1}^{M(t)} Z_n(t) \right) \qquad [2.33]$$

It can be shown that *when the system is closed for $n \geq n^*$*, i.e. when supernuclei are not created or consumed by other processes than attachment and detachment of monomers, the nucleation rate can be expressed as (see Appendix 1):

$$J(t) = j^*(t) - \frac{dn^*}{dt} Z^*(t) \qquad [2.34]$$

The overall nucleation rate can be expressed as a function of the characteristics of the nucleus only: $j^*(t)$ the flux of clusters "crossing" the critical size n^* and $Z^*(t)$ the concentration of nuclei.

2.2.4. Stationary nucleation

The resolution of stationary nucleation is probably the first achievement of the classical theory of nucleation. Accordingly, as will be shown in the following sections, stationary nucleation provides a reference framework for many non-stationary situations. The particular stationary regime described here is based on the following assumptions:

1) supersaturation and temperature are constant, hence the attachment and detachment frequencies are time independent;

2) the system is closed ($\forall n, K_n(t) = L_n(t) = 0$);

3) $\forall n \in [1; M - 1], j_n = F_n Z_n - B_{n+1} Z_{n+1} = J_s = $ constant.

A first observation is that owing to condition 1, the critical size is fixed (through equation [2.32]), therefore $\frac{dn^*}{dt} = 0$. Hence, equation [2.34] leads to:

$$J(t) = J_s \qquad [2.35]$$

Therefore, as expected for stationary nucleation, the overall nucleation rate is constant and equal to J_s.

Condition 3 corresponds to a set of $M - 1$ equations containing $M + 1$ unknowns ($Z_1 ... Z_M, J_s$). Therefore, two additional relations, corresponding to boundary conditions, should be added to solve this system, as discussed by McDonald [MCD 63].

The first condition is:

$$Z_M = 0 \qquad [2.36]$$

This means that the formation of an M-sized cluster is ruled out. This is acceptable, since this models aims at describing the system during nucleation, long before macroscopic domains are built.

The second condition stands:

$$Z_1 = C_1 \qquad\qquad [2.37]$$

The population of monomers in the old phase is set to its equilibrium value. This is in clear contradiction with condition 2: in a closed system, the monomers consumed in the nucleation process are not replenished, and their quantity should decrease. However, as both Z_1 and C_1 are large numbers, this approximation is acceptable [MCD 63].

The resolution of this system leads to (see Appendix 2, section A2.1 for details):

$$J_s = \left[\sum_{m=1}^{M-1} \frac{1}{F_m C_m} \right]^{-1} \qquad\qquad [2.38]$$

$$Z_n = C_n J_s \sum_{m=n}^{M-1} \frac{1}{F_m C_m} \qquad\qquad [2.39]$$

Therefore, the nucleation rate J_s and the cluster concentration distribution Z_n can be readily computed from the attachment frequencies F_n and the "equilibrium" concentrations C_n. As shown in equation [2.26], the latter are only functions of the work of cluster formation W_n. Hence, the knowledge of both attachment frequencies and work of cluster formation is central to further investigate the nucleation phenomenon. The determination of the former is is discussed in section 2.3. The evaluation of the work of cluster formation is the main topic of the following section.

2.2.5. Classical description of clusters: the capillarity approximation

Historically, the clusters have been described as small domains (or droplets) of the new phase within the old phase. As we shall see, this description relies on conventional thermodynamics. For this reason, *this approach is only accurate if the nucleus is large enough*, so it can be assimilated to an homogeneous phase. In this case, it is easier to consider n as *a continuous variable*, by contrast with the previous developments in which n was a discrete quantity. Indeed, all the relations derived in the previous section are easily transformed to the case where n is continuous, as shown in Table 2.1.

Flux	$j_n(t) = F_n(t)C_n(t)\left(\dfrac{Z_n(t)}{C_n(t)} - \dfrac{Z_{n+1}(t)}{C_{n+1}(t)}\right)$	[2.15]
	$j(n,t) = -F(n,t)C(n,t)\dfrac{\partial}{\partial n}\left(\dfrac{Z(n,t)}{C(n,t)}\right)$	[2.40]
Cluster concentration	$\dfrac{dZ_n(t)}{dt} = j_{n-1}(t) - j_n(t) + K_n(t) - L_n(t)$	[2.14]
	$\dfrac{\partial Z(n,t)}{\partial t} = -\dfrac{\partial}{\partial n}j(n,t) + K(n,t) - L(n,t)$	[2.41]
Nucleation rate	$J(t) = \dfrac{d}{dt}\left(\displaystyle\sum_{n^*(t)+1}^{M(t)} Z_n(t)\right)$	[2.33]
	$J(t) = \dfrac{d}{dt}\left(\displaystyle\int_{n^*(t)}^{M(t)} Z(n,t)\,dn\right)$	[2.42]
Equilibrium concentration	$C_n(t) = \dfrac{F_1(t)F_2(t)\cdots F_{n-1}(t)}{B_2(t)B_3(t)\cdots B_n(t)}C_1(t)$	[2.19]
	$C(n,t) = \exp\left[\displaystyle\int_1^n \ln\left(\dfrac{F(m,t)}{B(m,t)}\right)dm\right]C(1,t)$	[2.43]
Detachment frequencies	$B_n(t) = F_{n-1}(t)\exp\left(\dfrac{W(n,t)-W(n-1,t)}{kT}\right)$	[2.30]
	$B(n) = F(n)\exp\left(\dfrac{1}{kT}\dfrac{dW(n)}{dn}\right)$	[2.44]
Work for cluster formation	$W(n,t) = W(1,t) - kT(t)\displaystyle\sum_{m=1}^{n-1}\ln\left(\dfrac{F_m(t)}{B_{m+1}(t)}\right)$	[2.31]
	$W(n,t) = W(1,t) - kT(t)\displaystyle\int_1^n \ln\left(\dfrac{F_m(t)}{B_m(t)}\right)dm$	[2.45]

Table 2.1. *Compared useful expressions when n is considered as a continuous instead of a discrete variable*

2.2.5.1. The cluster's variables of state

To determine the cluster's variables of state, we first need to precisely define its boundaries with the old phase. This question may sound trivial, but it is not, even in the case of the nucleation of a solid phase. Indeed, there are many possibilities to place the surface that delimits the cluster from the old phase. As explained in Appendix 3, the position of the interface between the old phase and the cluster modify the composition of the latter by adding (or retrieving) n_i excess species. Therefore, the choice of this surface is not without consequences[5].

Once the cluster is clearly defined, its variables of state can be determined. Indeed, this tiny phase has its own properties, which are obviously different from the old phase, but which also differ from the new phase, due to its small size. Hence, the cluster has its own volume V_n, its own internal pressure P_n and each of the n monomers forming the cluster are considered to have the same chemical potential $\mu_{new,n}$. The latter is given by [GUG 85]:

$$\mu_{new,n}(P_n) = \mu_{new}(P) + \frac{1}{n} \int_P^{P_n} V_n(p)\,dp \qquad [2.46]$$

Of course, all these variables are n dependent and n is fixed by the choice of the dividing interface between the cluster and the old phase (according to Gibbs, see Appendix 3).

2.2.5.2. Energy balance of the cluster formation

Having defined the cluster and its variables of states, it is now possible to determine the energy balance for its formation.

For that purpose, let us consider the system in its most general initial state, i.e. when the old phase (containing M monomers) coexists with a substrate. The Gibbs energy of the system reads:

$$G_{old} = M\mu_{old} + \varphi_s(0) \qquad [2.47]$$

where the first summand is the Gibbs energy of M species in the old phase and $\varphi_s(0)$ is the interfacial energy between the substrate and the old phase. After

---

5 It is worth noting that the choice of this interface is of no consequence as far as equation [2.10] is concerned.

the formation of a cluster of the new phase, the Helmoltz free energy of the system is given by:

$$F_{inter} = [(M - n - n_i)\mu_{old} - PV_{M-n}] + [n\mu_{new,n} - P_n V_n]$$
$$+ n_i \mu_i + \varphi(n) + \varphi_s(n) \qquad [2.48]$$

where the various summands account for, respectively:

– the elimination of $n + n_i$ species from the old phase;

– the formation of a cluster of n species;

– the presence of n_i excess species at the cluster/old phase interface (see Appendix 3);

– the interfacial energy $\varphi(n)$ between the cluster and its environment (old phase but also substrate in the case of heterogeneous nucleation – this energy also depends on the choice of the dividing surface to define the cluster);

– and finally the interfacial energy $\varphi_s(n)$ of the remaining interface between the old phase and the substrate.

Obviously, this description is suited to study both heterogeneous ($\varphi_s(n) < \varphi_s(0)$) and homogeneous ($\varphi_s(n) = \varphi_s(0)$) nucleations. In all cases, $\varphi_s(0)$ is given by:

$$\varphi_s(0) = \sigma_i A_s \qquad [2.49]$$

where A_s and σ_i are the surface area and specific energy of the substrate/old phase interface, respectively. The Helmoltz free energy relates to the Gibbs energy [GUG 85]:

$$\begin{aligned} G_{inter} &= F_{inter} + PV \\ &= (M - n - n_i)\mu_{old} + n\mu_{new,n} - (P_n - P)V_n \qquad [2.50] \\ &\quad + n_i \mu_i + \varphi(n) + \varphi_s(n) \end{aligned}$$

Combined with equations [2.6], [2.7] and [2.46], the work of formation of the cluster can be expressed as:

$$W(n) = n\Delta\mu + \int_P^{P_n} V_n(p)\,\mathrm{d}p - (P_n - P)V_n + n_i(\mu_i - \mu_{old})$$
$$+ \varphi(n) + \varphi_s(n) - \varphi_s(0) \qquad [2.51]$$

which indeed corresponds to the general form of equation [2.8].

Equation [2.51] is a very general expression for the work of formation of a cluster. Its sole limitation is that it only concerns one type of species (see section 2.1.2). However, this expression is not easy to handle for practical purposes. Therefore, further assumptions are usually made to simplify it.

2.2.5.3. The capillarity approximation

The first simplification consists of defining the cluster using the *equimolar surface* (see Appendix 3). In this case, the excess population of n_i interfacial species vanishes ($n_i = 0$).

Considering *condensed phases* brings further simplification. Indeed, a condensed cluster is essentially incompressible, so each of its n constituting monomers occupies the same volume v_0, so that:

$$V_n(P_n) = nv_0 \qquad [2.52]$$

In addition, all the pressure-dependent terms vanish, and equation [2.51] becomes:

$$W(n) = n\Delta\mu + \varphi(n) + \varphi_s(n) - \varphi_s(0) \qquad [2.53]$$

Finally, the surface energy $\varphi(n)$ is assumed to scale with S_n the surface area between the cluster and the old phase, and $S_{n,s}$ the (optional) surface area between the cluster and the substrate:

$$\varphi(n) = \sigma_n S_n + \sigma_{n,s} S_{n,s} \qquad [2.54]$$

where the coefficients σ_n and $\sigma_{n,s}$ are the specific surface energies of the two interfaces, respectively.

Equation [2.54] is a central assumption in the classical theory of nucleation. It describes the cluster as a "droplet" of the new phase in the old phase. The energetic penalty to overcome during nucleation is associated with surface energies as used in Young's law, see equation [2.62]. Therefore, it is often referred to as a capillarity approximation. In addition, in the classical theory of nucleation, these specific surface energies are considered to be independent on the size of the cluster (i.e. $\sigma_n = \sigma$ and $\sigma_{n,s} = \sigma_s$).

2.2.5.4. Homogeneous nucleation

In the case of homogeneous nucleation, the cluster is often assumed to adopt a *regular shape*. In this case, the surface energy $\varphi(n)$ becomes:

$$\varphi(n) = \sigma c V_n^{2/3} = \sigma c (nv_0)^{2/3} \qquad [2.55]$$

where c is a shape factor (c is n independent, and $c = (36\pi)^{1/3}$ for a sphere). The work of formation of the cluster is simplified into:

$$W(n) = n\Delta\mu + \sigma c (nv_0)^{2/3} \qquad [2.56]$$

Now, the application of the maximization condition of $W(n)$ equation [2.9] determines the size of the nucleus. In the case of the homogeneous nucleation of regularly shaped clusters, the critical size is given by:

$$n^* = \frac{8\sigma^3 c^3 v_0^2}{27(-\Delta\mu)^3} \qquad [2.57]$$

And the corresponding work of nucleation reads:

$$W^* = \frac{4\sigma^3 c^3 v_0^2}{27(-\Delta\mu)^2} \qquad [2.58]$$

Interestingly enough, according to both formulae, it is found that $W^* = -\frac{1}{2}n^*\Delta\mu$ in agreement with the discussion in section 2.1.3.

2.2.5.5. Heterogeneous nucleation

Another interesting example is the heterogeneous nucleation of a cluster with the shape of a truncated sphere of radius R, characterized by its wetting angle θ with the substrate. In this case, the spherical surface S_R and the base $S_{R,s}$ are given by:

$$S_R = 2\pi R^2 (1 - \cos\theta) \quad \text{and} \quad S_{R,s} = \pi R^2 \sin^2\theta \qquad [2.59]$$

The volume of the cluster is given by:

$$V_R = \Psi(\theta)\frac{4}{3}\pi R^3 \quad \text{with} \quad \Psi(\theta) = \frac{1}{4}(2 + \cos\theta)(1 - \cos\theta)^2 < 1 \qquad [2.60]$$

The work of formation of such a cluster is given by equation [2.53]. The relation between n and V_R (equation [2.52]) yields:

$$W(R) = \Psi(\theta)\frac{4}{3}\pi R^3 \frac{1}{v_0}\Delta\mu + \sigma 2\pi R^2(1 - \cos\theta) + (\sigma_i - \sigma_s)\pi R^2 \sin^2\theta$$

$$[2.61]$$

As θ is related to the interfacial specific energies according to Young's law:

$$\cos\theta = \frac{\sigma_i - \sigma_s}{\sigma} \qquad [2.62]$$

the following expression is derived:

$$W(R) = \Psi(\theta)\left(\frac{4}{3}\pi R^3 \frac{1}{v_0}\Delta\mu + 4\sigma\pi R^2\right) \qquad [2.63]$$

This work can be expressed as a function of n using equation [2.52] to yield:

$$W(n) = n\Delta\mu + \Psi(\theta)^{1/3}\sigma c(nv_0)^{2/3} \qquad [2.64]$$

This equation is similar to equation [2.56] in the case of homogeneous nucleation, as the work of cluster formation only differs by a factor of $\Psi^{1/3}$.

Finally, the size and work of formation of the nucleus are given by:

$$n^* = \Psi(\theta)\frac{8\sigma^3 c^3 v_0^2}{27(-\Delta\mu)^3} \quad \text{and} \quad W^* = \Psi(\theta)\frac{4\sigma^3 c^3 v_0^2}{27(-\Delta\mu)^2} \qquad [2.65]$$

By comparison with homogeneous nucleation, both quantities are reduced by a Ψ factor (it is reminded that $\Psi < 1$). It is also easy to establish that the radius R of the truncated sphere is the same as the radius of the corresponding spherical nucleus in homogeneous nucleation (Figure 2.4).

It thus becomes obvious that heterogeneous nucleation is easier than homogeneous nucleation. Moreover, the better the cluster wets the substrate surface – i.e. the smaller the contact angle θ – the more favored will be the heterogeneous over homogeneous nucleation.

Homogeneous   Heterogeneous

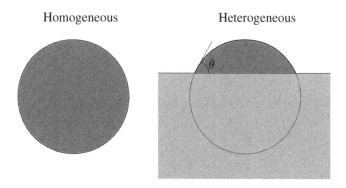

Figure 2.4. *Comparison of the spherical nuclei in homogeneous and heterogeneous nucleation*

2.2.5.6. The quadratic approximation

In summary, for homogeneous nucleation of regularly shaped clusters or heterogeneous nucleation of a spherical cap, the work of cluster formation is given by the general formula:

$$W(n) = n\Delta\mu + 3\alpha n^{2/3} \qquad [2.66]$$

with

$$\alpha = \frac{\left(\sigma^3 c^3 v_0^2\right)^{1/3}}{3} \qquad [2.67]$$

$$\alpha = \frac{\left(\psi(\theta)\sigma^3(36\pi)v_0^2\right)^{1/3}}{3} \qquad [2.68]$$

for homogeneous and heterogeneous cases, respectively.

This expression of $W(n)$ was the missing information needed to solve the system of equations describing stationary nucleation in section 2.2.4. This resolution is usually carried out by approximating $W(n)$ in the vicinity of n^* by the truncated Taylor development:

$$W(n) \simeq W^* + \frac{1}{2}\left(\frac{d^2W}{dn^2}\right)_{n=n^*}(n-n^*)^2 \qquad [2.69]$$

This approximation is sometimes referred to as the *quadratic approximation*.

It is interesting to consider the particular region around n^* called the *nucleus region* where the variation of $W(n)$ does not exceed kT (Figure 2.5). According to equation [2.69], this region is limited by:

$$W(n) - W^* < kT \Leftrightarrow |n-n^*| < \frac{1}{\beta} \qquad [2.70]$$

where β is defined by:

$$\beta = \left[\frac{1}{2kT}\left(-\frac{d^2W}{dn^2}\right)_{n=n^*}\right]^{1/2} \qquad [2.71]$$

It follows that:

$$\beta = \left[\frac{1}{kT}\frac{\alpha}{3}\frac{1}{(n^*)^{4/3}}\right]^{1/2} \qquad [2.72]$$

From equations [2.57], [2.58] and [2.65], the following expressions are found for n^* and W^*:

$$n^* = \frac{8\alpha^3}{(-\Delta\mu)^3} \qquad [2.73]$$

$$W^* = \frac{4\alpha^3}{(-\Delta\mu)^2} \qquad [2.74]$$

leading to:

$$\beta = \left[\frac{1}{kT}\frac{1}{3}\frac{W^*}{(n^*)^2}\right]^{1/2} = \left[\frac{1}{kT}\frac{1}{6}\frac{-\Delta\mu}{n^*}\right]^{1/2} \qquad [2.75]$$

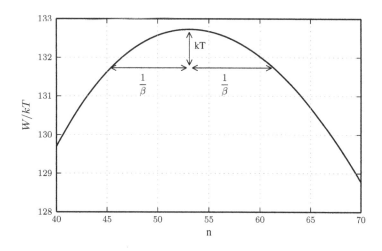

Figure 2.5. *Evolution of cluster work of formation with size* n, *as computed from equation [2.56] (homogeneous nucleation of spherical clusters with the following values:* $T = 300\,\mathrm{K}$, $\Delta\mu = -5\,\mathrm{kT}$, $\sigma = 0.5\,\mathrm{Jm^{-2}}$ *and* $v_0 = 1 \times 10^{-29}\,\mathrm{m^3}$. *The critical size is* $n^* = 53$, *and the work of nucleation is* $W^* = 132.7\,\mathrm{kT}$.). *The nucleus region is defined by equation [2.70] in which all clusters are energetically equivalent to the nucleus*

Since typically $\frac{-\Delta\mu}{kT} < 10$, usually $\beta \lesssim 1$.

All clusters within the nucleus region can be considered as energetically equivalent, as their work of formation differs from less than kT. *Accordingly, and following equation [2.26], the distribution of this population is accurately described in this size domain by:*

$$C(n) \simeq C(0)\exp\left[-\frac{W^*}{kT}\right]\exp\left[\beta^2(n-n^*)^2\right]$$
$$= C^*\exp\left[\beta^2(n-n^*)^2\right] \qquad [2.76]$$

2.2.5.7. Stationary nucleation in the capillarity approximation

Now, we are ready to go back to the description of stationary nucleation mentioned in section 2.2.4. Because n is now considered as a continuous quantity, the the discrete system of equations [2.38] and [2.39] is translated into (see Appendix 2, section A2.2 for details):

$$Z(n) = C(n) \left[\int_1^M \frac{1}{F(m)C(m)} \, dm \right]^{-1} \int_n^M \frac{1}{F(m)C(m)} \, dm$$

[2.77]

$$J_s = \left[\int_1^M \frac{1}{F(m)C(m)} \, dm \right]^{-1}$$

[2.78]

Using the quadratic approximation (section 2.2.5.6), the integrals in the latter expressions may be approximated under the following assumptions:

– $C(n)$ is approximated by equation [2.76];

– $F(n)$ is approximated by F^*;

– M is replaced by $+\infty$.

Then, by definition of the error function[6]:

$$\int_a^M \frac{1}{F_m C_m} \, dm \simeq \frac{1}{F^* C^*} \int_a^{+\infty} \exp\left[-\beta^2 (m - n^*)^2 \right] \, dm$$
$$= \frac{\sqrt{\pi} \left[1 - \mathrm{erf}\left(\beta(a - n^*) \right) \right]}{2\beta F^* C^*}$$

[2.79]

Injecting these approximations in equation [2.77] yields:

$$Z(n) \simeq C(n) \frac{1 - \mathrm{erf}\left[\beta(n - n^*) \right]}{1 - \mathrm{erf}\left[\beta(1 - n^*) \right]}$$

[2.80]

---

6 By definition,

$$\mathrm{erfc}(x) = \frac{2}{\sqrt{\pi}} \int_x^\infty \exp\left(-t2 \right) \, dt = 1 - \mathrm{erf}(x)$$

where

$$\mathrm{erf}(x) = \frac{2}{\sqrt{\pi}} \int_0^x \exp\left(-t2 \right) \, dt.$$

Under the condition $(n^* - 1) > 1/\beta$ [6], $\mathrm{erf}\,[\beta(1 - n^*)] \simeq \mathrm{erf}(-\infty) = -1$ and this expression can be further simplified into:

$$Z(n) \simeq \frac{1}{2}C(n)\,(1 - \mathrm{erf}\,[\beta(n - n^*)])$$ [2.81]

Similarly, the expression of J_s is transformed into:

$$J_s \simeq zF^*C^*$$ [2.82]

where

$$z = \frac{2\beta}{\sqrt{\pi}\,(1 - erf\,[\beta(1 - n^*)])}$$ [2.83]

is known as the *Zeldovich factor*. Under the condition $(n^* - 1) > 1/\beta$, this factor assumes the simple form:

$$z = \frac{2\beta}{\sqrt{\pi}}$$ [2.84]

From equation [2.82], it is clear that this factor mitigates the flux of clusters "crossing" the critical size (just as β, $z < 1$, see section 2.2.5.6). Physically, and based on the fact that all clusters are energetically equivalent in the nucleus region (see section 2.2.5.6), a subnucleus reaching a critical (or near-critical) size does not necessarily evolve into a supernucleus, as inferred from purely thermodynamic considerations. Instead, because of thermally activated random detachment processes, it can return back into the subcritical size domain (only clusters with sizes of $n^* + \frac{2}{z\sqrt{\pi}}$ or more become for sure stable supernuclei). Replacing C^* in equation [2.82] by its expression equation [2.26], the famous relation is found:

---

6 Considering the definition of β (equation [2.75]) and approximating $n^* - 1 = n^*$, this condition is equivalent to $W^* > 3kT$.

$$J_s \simeq zF^*C(0)\exp\left[-\frac{W^*}{kT}\right]$$ [2.85]

This expression states that the stationary nucleation rate exponentially depends on the nucleation work. The pre-exponential factor is the product of:

– the probability that a nucleus evolves into a supernucleus (z);

– the concentration of nucleation sites ($C(0)$, see section 2.2.6.2);

– the attachment frequency of a monomer to the nucleus (F^*).

2.2.6. *Time-dependent nucleation*

So far, only the situations in which the system was under equilibrium or steady state have been examined. In this section, cases of time-dependent evolutions of the system will be surveyed in the framework of the classical nucleation theory. This time dependence may arise from several causes. However, in each case depicted in the following, the evolution of the system is always described in the light of the steady state discussed in section 2.2.5.7.

2.2.6.1. *Transient nucleation at constant supersaturation*

The closest to the previously discussed case of stationary nucleation where time dependence of the nucleation can be observed is when *nucleation starts*. Indeed, upon establishment of the supersaturation, the system requires an *incubation time*, also referred to as *time lag*, before the stationary nucleation regime is reached.

In its most simple expression, this transient regime is described under the following assumptions:

– the system is closed: $\forall n, K_n(t) = L_n(t) = 0$;

– the supersaturation is abruptly established at $t = 0$ to a constant value (i.e. the supersaturation follows a staircase-like time evolution);

– at $t = 0$, there is no cluster of the new phase: $\forall n, \quad Z(n, 0) = 0$.

From these conditions, it is clear that the asymptotic state of the system is the stationary nucleation described in section 2.2.5.7. But before this steady-state is reached, the nucleation rate has to progressively increase from

0 to J_s. To describe this transient, it is possible to solve the differential equations [2.40] and [2.41] in Table 2.1 and determine $Z(n,t)$ and $J(t)$ (with the help of equation [2.34]). This resolution, which is beyond the scope of this work, is usually carried out using the boundary conditions in section 2.2.4, in the nucleus size region, under the parabolic approximation of the work of nucleation equation [2.69] used in section 2.2.5.7. In this description, the time lag τ of nucleation is expressed as [KAS 00]:

$$\tau = \frac{4}{\pi^2 \beta^2 F^*} \tag{2.86}$$

In a slightly different approach, the *delay time* of nucleation can be defined from the integral form of equation [2.42]:

$$\int_0^t J(t)\,dt = \int_{n^*(t)}^{M(t)} Z(n,t)\,dn = N(t) \tag{2.87}$$

where $N(t)$ is the concentration of supernuclei present in the system at time t (it is recalled that under the assumptions considered here, $N(0) = 0$). The asymptotic establishment of the stationary nucleation regime can be described by Figure 2.6:

$$N(t) = J_s(t - t_i) + \epsilon(t) \quad \text{with} \quad \lim_{t \to +\infty} (\epsilon(t)) = 0 \tag{2.88}$$

where t_i is the delay time of nucleation. The following expression of t_i was derived by Shneidman and Weinberg [SHN 92] under the parabolic approximation for nucleation work:

$$t_i = \left(\frac{\gamma}{2} + \ln(\beta n^*)\right) \frac{\pi^2}{8}\tau \tag{2.89}$$

where $\gamma = 0.5772$ is Euler's constant, and $\beta n^* = \frac{W^*}{3kT}$ (according to equation [2.75]). Therefore, t_i is proportional to (and larger than) τ.

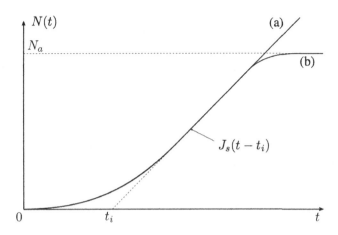

Figure 2.6. *Schematic representation of the time evolution of supernuclei population during transient nucleation a) without and b) with a limited number of nucleation sites*

Upon abrupt establishment of supersaturation, the system requires a delay time before the stationary nucleation regime corresponding to this supersaturation is reached. There exist several ways to quantify the associated delay time, but all the definitions are proportional to τ as defined in equation [2.86].

2.2.6.2. Nucleation on a limited number of active sites

Another cause for non-stationary nucleation examined here is the existence of a limited number of *active nucleation sites*. Indeed, in section 2.2.5.7, it was shown that the nucleation rate scales with $C(0)$, which represents the concentration of nucleation sites (see equation [2.85]).

In heterogeneous nucleation, these sites are usually defects on the surface of the substrate, such as crystallographic steps or kinks, chemical impurities or adsorbed foreign particles. When nucleation occurs in the volume of the old phase, similar defects may act as nucleation centers. If these defects are chemical impurities or particles, it is not rigorous to speak of homogeneous nucleation. However, for the sake of brevity, we shall use this term in this section to refer to this situation.

Let us again consider the situation in which:

– the system is closed: $\forall n, K(n, t) - L(n, t) = 0$;

– the supersaturation is abruptly established at $t = 0$ to a constant value;

– at $t = 0$, there is no cluster of the new phase: $\forall n, Z(n, 0) = 0$;

– the active centers have the same nucleation activity.

It has been shown earlier that under these assumptions, the nucleation rate $J(t)$ progressively increases to reach its stationary value J_s (see Figure 2.6). In the situation considered here, the nucleation rate decreases with the number of remaining nucleation sites through:

$$J(t) = J_a(t) \left(N_a - N(t) \right)$$ [2.90]

where N_a is the initial concentration of active sites ($C(0) = N_a$), and $N(t)$ is the concentration of supernuclei formed at time t. J_a thus represents the nucleation rate per nucleation site. The application of equation [2.42] leads to:

$$\frac{dN(t)}{dt} = J_a(t) \left(N_a - N(t) \right)$$ [2.91]

The solution of this equation reads (it is recalled that $N(0) = 0$):

$$N(t) = N_a \left(1 - \exp \left[- \int_0^t J_a(t') \, dt' \right] \right)$$ [2.92]

The resulting evolution of $N(t)$ is schematically plotted in Figure 2.6. By comparison with Figure 2.6, the concentration of supernuclei saturates at N_a and the nucleation rate drops to 0.

2.2.7. Limitations of the classical theory of nucleation

In conclusion, the classical theory of nucleation based on the capillarity description of the nucleus is attractive, as it reduces the complex system of equations describing nucleation into a rather simple set of equations. However, this simplification relies on a series of hypotheses that limit its range of application.

A first question is the practical determination of the specific surface tensions used to compute the energetic cost of formation of the cluster/old phase and cluster/substrate interfaces. Unfortunately, these specific surface tensions are not known *a priori*, since they can deviate to a large extent from the values that can be measured experimentally at a macroscopic scale between the (bulk) new phase and the (bulk) old phase or substrate.

This brings us to a more profound limitation of this thermodynamic treatment of nucleation. In this approach, macroscopic-like and homogeneous thermodynamic properties are used to describe the cluster, even if they may deviate from the "bulk" quantities (section 2.2.5.1). In fact, such a description should only be applied to large enough clusters, containing a large enough number of species. From the examples of homogeneous and heterogeneous nucleation (sections 2.2.5.4 and 2.2.5.5), the number n^* of species constituting the nucleus is found to vary as $1/\Delta\mu^3$. Therefore, as supersaturation increases, the size of the nucleus rapidly decreases down to sizes where $n^* < 100$. In such a situation, the "average" description of volume or surface properties of the clusters proposed in section 2.2.5.1 is not rigorous.

Finally, another weakness of the theoretical treatment above is that the entropic contribution of translation and rotation motions of the clusters have not been taken into account. This contribution has been introduced by Lothe and Pound [LOT 62].

All these considerations have led to more sophisticated (and accurate) theoretical descriptions of the cluster, as recently discussed in great detail by Mutaftschiev [MUT 13]. However, for the practical cases discussed in the following chapters, the present description is sufficient to introduce key concepts such as

– the difference between homogeneous and heterogeneous nucleation;

– the existence of a critical cluster or nucleus;

– the stationary nucleation rate and its dependence to the nucleation work;

– the time lag in nucleation.

2.3. Growth

Growth is the mechanism by which the new phase absorbs monomers from the old phase. As soon as nuclei are formed, they start growing and eventually coalesce. The formalism associated with the latter process will be described in section 2.5. In the late stages of phase transformation, macroscopic domains of the new phase also progress through this mechanism, as will be shown in section 2.3.2.

Importantly, the attachment of monomers from the old phase is also involved in the nucleation step as explained in section 2.2.1. In fact, the knowledge of the attachment frequencies is central in the determination of the nucleation rates (for instance the attachment frequency of a monomer to the nucleus is needed to compute the nucleation rate in equation [2.85]).

2.3.1. *Attachment of monomers*

The transition frequency F_n (s^{-1}) introduced in section 2.2.1 corresponds to the probability that monomers attach to an n-sized cluster. More generally, this probability scales with:

– the external surface S_n of the cluster;

– the *local* concentration of monomers;

– the actual frequency of attachment, i.e. the probability that a *close enough* monomer actually sticks to the cluster.

Naturally, the latter depends on the interaction involved between the monomer and the cluster. For instance, the attachment may proceed by ballistic collision, or by incorporation into the crystal lattice of the cluster of already adsorbed species. Also, based on this description, phenomena such as diffusion of monomers from within the old phase may be accounted for directly in the formulation of F_n. In this context, the notions of *local* or *close enough* are not well defined and depend on the actual mechanism considered for monomer attachment. Owing to the fact that this work's main focus is the solidification from solutions, and also to clearly separate mass transport phenomena from actual attachment and detachment processes, the following approach is proposed here without much loss of generality.

It is considered that only monomers already adsorbed on the clusters may be incorporated into them. Therefore, the mechanism for monomer attachment is the incorporation of adsorbates (such as adatoms). These species are thermally mobile on the cluster's surface, as described by Frenkel [FRE 55]. This movement proceeds by successive jumps over a distance comparable to the size of the adsorbed monomers. For each jump, the adsorbate has to overcome an energy barrier $E_{j,n} > 0$. Hence, the frequency of such an event is given by:

$$\nu_{j,n} = \nu_n \exp\left(\frac{-E_{j,n}}{kT}\right) \qquad [2.93]$$

where ν_n is the vibration frequency of an adsorbed monomer on an n-sized cluster. Hence, from the considerations above, this specific mechanism of attachment leads to:

$$F_n = \gamma_n \nu_{j,n} c_{ad} S_n \qquad [2.94]$$

where γ_n, the *sticking coefficient*, is the probability that a jump actually results in the incorporation of a monomer. In this equation, c_{ad}, the surface concentration of adsorbed monomers, is supposed not to be n dependent, and is closely related to Z_1, the monomer concentration in the old phase. As shown by [FRE 55], the diffusion coefficient D_{ad} of the adsorbate is directly related to $\nu_{j,n}$ by:

$$D_{ad,n} = d_{ad}^2 \nu_{j,n} \qquad [2.95]$$

where d_{ad} is the diameter of the adsorbed monomer. Therefore, the transition frequency F_n becomes:

$$F_n = \gamma_n \frac{D_{ad,n}}{d_{ad}^2} c_{ad} S_n \qquad [2.96]$$

Finally, it must be noted that this equation is generally simplified by neglecting the n dependence of γ_n, $E_{j,n}$ and ν_n. Therefore, $D_{ad,n} = D_{ad} = $ constant, and:

$$F_n = \gamma \frac{D_{ad}}{d_{ad}^2} c_{ad} S_n = F_{ad} S_n \qquad [2.97]$$

2.3.2. Crystal growth

Let us now examine how supernuclei grow. In this section, we will also anticipate the case of the growth of macroscopic crystals, which is the main mechanism by which the new phase progresses in the late stages of phase transformation.

For this purpose, let us consider a domain of the new phase, characterized by its size $m \gg n^*$. By definition, the surface S_m of contact between this domain and the old phase is large in the case of a macroscopic domain. The growth rate g_c measures the progression of the new phase normally to this surface. This growth may proceed through various mechanisms for large crystals, depending on their surface characteristics [KAS 00]:

– *layer-by-layer growth*: if the surface is smooth at the molecular level, the growth may proceed by nucleation of 2D domains, which grow to fully cover the surface. The same process iterates on the so formed surface;

– *island nucleation and growth*: this case corresponds to the heterogeneous nucleation described in section 2.2.5.5, except that in this case the substrate is the m-sized domain of the new phase. This is an intermediate situation, often encountered in practical cases;

– *continuous growth*: this case corresponds to a rough surface at the molecular level. By contrast with the previous cases, every site of the surface may attach a monomer and participate to the growth.

The two first cases are restricted to macroscopic domains and may be described on the basis of the nucleation theory developed in this chapter. For the sake of brevity, we shall thus restrict ourselves to the case of continuous growth, which is the limiting regime of the nucleation-mediated growth of macroscopic domains[7], and is also relevant to describe the growth of supernuclei. It also corresponds to the fastest growth rate. This situation is the same as described in sections 2.2.1 and 2.3.1: the growth proceeds by attachment (and detachment) of monomers to (respectively, from) lattice sites of the crystal surface. In this case, the growth rate is given by:

$$g_c = d_{ad}(F_m - B_m) \qquad [2.98]$$

assuming that an attached (respectively, detached) monomer displaces the surface of the new phase by a distance d_{ad} (which is the diameter of a

---

7 When growth proceeds by nucleation of islands, the surface becomes rougher, and more active sites for growth are created.

monomer). The expression of F_m is given by equation [2.97]. According to equation [2.30], B_m is given by:

$$B_m = F_{m-1} \exp \left(\frac{W(m) - W(m-1)}{kT} \right)$$ [2.99]

Using equation [2.8], this leads to

$$B_m = F_{m-1} \exp \left[\frac{1}{kT} \left(m\Delta\mu + \Phi_m - (m-1)\Delta\mu - \Phi_{m-1} \right) \right]$$ [2.100]

If m is large enough (which is true at least for macroscopic domains), it is reasonable to consider that $S_{m-1} \simeq S_m$. Considering equation [2.97], this leads to: $F_{m-1} \simeq F_m$. Also, the difference $\Phi_m - \Phi_{m-1}$ is negligible. Hence:

$$B_m = F_m \exp \left(\frac{\Delta\mu}{kT} \right)$$ [2.101]

Therefore, equation [2.98] leads to:

$$g_c = d_{ad} F_m \left[1 - \exp \left(\frac{\Delta\mu}{kT} \right) \right]$$ [2.102]

where F_m may be considered as a constant (if in addition to S_m the variations of c_{ad} may be neglected). In this case, we find that for small supersaturation values:

$$g_c \simeq d_{ad} F_m \frac{\Delta\mu}{kT} \propto \Delta\mu$$ [2.103]

If supersaturation is constant, the growth rate is constant as well.

2.4. Aggregation

So far, we have treated the nucleation and growth of isolated nuclei. The fact that nuclei do not interact was indeed a basic assumption for the development of the nucleation theory in section 2.2.1. This hypothesis is probably valid in the early stages of phase transformation, but is certainly not in the later stages. Therefore, a new formalism has to be set up to describe the progression of the new phase well beyond nucleation. This is the purpose of this and the following sections.

The progression of the new phase may include a variety of concomitant phenomena that may not be easy to separate in practical studies or even conceptually. For instance, what is the difference between the *aggregation* and the *coalescence* of supernuclei? In both cases, two clusters meet to form a larger domain of the new phase.

Here, we define the aggregation as the clustering of two mobile supernuclei of fixed sizes. This process will be discussed in this section. By contrast, coalescence would result from the contact between two growing supernuclei and is the focus of the following section.

2.4.1. General equations for phase transformation

In a first approach, it is possible to generalize the model discussed in section 2.2.1 to the case when not only monomers, but also larger clusters may attach to (or detach from) a given cluster of size n. This is done by considering that the transformation of an n-sized cluster into an m-sized cluster occurs at a frequency denoted as $F_{nm}(t)$[8] (again, this frequency may be time dependent). Of course, $\forall n, F_{nn}(t) = 0$. In this case, the net creation of new n-sized clusters is given by:

$$\left[\frac{dZ_n(t)}{dt} \right]^+ = \sum_{m=1}^{M(t)} F_{mn}(t) Z_m(t) + K_n(t) \qquad [2.104]$$

where the sum corresponds to the formation of n-sized clusters by transformation of existing m-sized clusters (with $m \neq n$, it is recalled that $\forall n, F_{nn}(t) = 0$), and the second term accounts for the apparition of n-sized clusters by other processes (open system).

---

8 We obviously find that $F_n(t) = F_{n(n+1)}(t)$ (aggregation of a monomer to an n-sized cluster, $n \geq 1$) and $B_n(t) = F_{n(n-1)}(t)$ (release of a monomer by an n-sized cluster, $n > 1$).

Symmetrically, n-sized clusters vanish following:

$$\left[\frac{dZ_n(t)}{dt}\right]^{-} = -\sum_{m=1}^{M(t)} F_{nm}(t)Z_n(t) - L_n(t) \qquad [2.105]$$

Consequently, the time evolution of the concentration of n-sized clusters $Z_n(t)$ corresponds to the balance of both contributions, leading to the generalized form of equation [2.11]:

$$\frac{dZ_n(t)}{dt} = \sum_{m=1}^{M(t)} (F_{mn}(t)Z_m(t) - F_{nm}(t)Z_n(t)) + K_n(t) - L_n(t) \qquad [2.106]$$

It is also possible to generalize the flux of clusters that grow or decay "across" a given size n. To express this flux, let us first consider the number of clusters whose size is $\leq n$ that grow into larger clusters with a specific size m ($m > n$) [KAS 00]. This number is shown in Figure 2.7:

$$j_{n,m}(t) = \sum_{m'=1}^{n} (F_{m'm}(t)Z_{m'}(t) - F_{mm'}(t)Z_m(t)) \qquad [2.107]$$

The global flux of clusters growing or decaying "through" size n is the sum of these $j_{n,m}(t)$ for all m values such as $m > n$ and is thus given by:

$$\begin{aligned} j_n(t) &= \sum_{m=n+1}^{M(t)} j_{n,m}(t) = \sum_{m=n+1}^{M(t)} \sum_{m'=1}^{n} (F_{m'm}(t)Z_{m'}(t) \\ &\quad - F_{mm'}(t)Z_m(t)) \end{aligned} \qquad [2.108]$$

Then:

$$j_{n-1}(t) - j_n(t) = \sum_{m=1}^{M(t)} (F_{mn}(t)Z_m(t) - F_{nm}(t)Z_n(t)) \qquad [2.109]$$

And equation [2.106] becomes:

$$\frac{dZ_n(t)}{dt} = j_{n-1}(t) - j_n(t) + K_n(t) - L_n(t) \qquad [2.110]$$

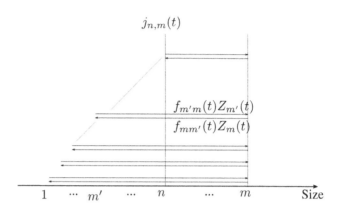

Figure 2.7. *Determination of the number of clusters whose size is $\leq n$ that grow into larger clusters with a specific size m ($m > n$). Adapted from [KAS 00]*

Interestingly enough, the system of M equations [2.106] fully describes the system and its evolution *at all stages of phase transformation*. However, as the system departs from nucleation and enters aggregation or coalescence stages, the exact resolution of this system becomes mathematically impractical. We will see in section 2.5 that a completely different formalism is more efficient at describing coalescence and associated phenomena. In the case of aggregation, the present approach is still usable with some simplifications.

2.4.2. *Aggregative growth*

Actual processes that lead to the formation of nanocrystals such as those described in section 3.2 and Chapter 6 have been described by some authors as a two-step process [WAN 14]. The first step would correspond to the classical nucleation exposed in the previous section. Under optimized conditions, this nucleation would provide small supernuclei just above the critical size[9].

These supernuclei would aggregate in a second step into larger particles. Interestingly enough, this aggregation process could be described as a secondary nucleation in which [WAN 14]:

---

9 This is possible through the Lamer mechanism depicted in section 3.2.1.

– the monomers are the supernuclei formed by primary nucleation;

– whose aggregation is driven by attractive interactions (such as Van der Walls forces);

– with an energetic penalty associated with repulsive interactions (for example, electrostatic or steric; see sections 3.2.2 and 6.3).

The mathematical description of this process can thus be achieved by the same formalism as in section 2.2.1, in which the equations are an other (but similar) subset of the more general system in section 2.4.1. In the most simple version of this new set of equations, monomers would have a size $n_{mono} > 1$, and cluster sizes would be multiples of n_{mono}. The generation of monomers by nucleation would be accounted for by $K_{n_{mono}}(t) = J(t)$, where $J(t)$ is the nucleation rate.

There are experimental evidences of this two-step process, the most striking being the emergence, at intermediate times of phase transformation, of bimodal size distributions of the clusters, corresponding to the co-existence of isolated "monomers" along with aggregated particles at (or near to) their final size. Therefore, such a mechanism cannot be discarded when examining nucleation and growth processes such as those involved in the fabrication of nanoparticles.

2.5. Coalescence

As briefly discussed in section 2.4.1, the kinetic model of phase transition is suitable to describe not only the nucleation, but also the coalescence and growth of the new phase. However, it is mathematically impractical. Therefore, this model cannot be actually used to investigate the late coalescence and growth stages. Hence, an alternative approach must be adopted, in which the temporal evolution of the overall transformed volume fraction is examined. This approach is best suited to describe the progression of immobile clusters (it has been developed in the first place to describe phase transformations in the solid state).

2.5.1. The Johnson–Mehl–Avrami–Kolmogorov equation

In section 2.2.6.2, the question of the nucleation over a limited number of nucleation sites has been treated, and the time evolution of the concentration of supernuclei could be determined. The purpose of this section is to extend this approach to more advanced stages of the phase transformation, when supernuclei can grow to such an extent that they may contact each other.

In equations [2.90] and [2.91], it was implicitly assumed that the nucleation itself is the only way by which available nucleation sites are consumed. If such an assumption is reasonable during early stages of nucleation, it becomes questionable when phase transformation progresses. Indeed, *nucleation sites may also be ingested by growing supernuclei*. Therefore, a more general relation must be derived. This can be done considering that:

– the system is infinite;

– the nucleation is homogeneous, and the clusters are spherical;

– the nucleation is a random and uniform process;

– cluster growth stops at contact points between adjacent supernuclei.

In this case, the transformed volume fraction is given by [WEI 97]:

$$\alpha(t) = 1 - \exp\left[-\frac{4\pi}{3}N_a \int_0^t J_a(t') \left(\int_{t'}^t \dot{R}(t'')\,dt''\right)^3 dt'\right] \quad [2.111]$$

where the growth rate $\dot{R}(t)$ is rigorously given by (considering the definition of $F(n)$ in equation [2.97])[10]:

$$\dot{R}(t) = F_{ad}v_0 \left\{1 - \exp\left[\frac{\Delta\mu}{kT}\left(1 - \left(\frac{n^*}{n}\right)^{1/3}\right)\right]\right\} \quad [2.112]$$

Equation [2.111] is known as the Johnson–Mehl–Avrami–Kolmogorov (JMAK) equation.

New supernuclei can only appear in untransformed regions, which represent a volume given by $V[1 - \alpha(t)]$. Therefore, equation [2.91] is modified into:

$$\frac{dN(t)}{dt} = J_a(t)N_a[1 - \alpha(t)] \quad [2.113]$$

---

10 A similar derivation is proposed in section 2.3, in the case of the late stages of phase transformation (corresponding to the limiting case where $n \gg n^*$).

which yields upon integration, assuming that $N(0) = 0$:

$$N(t) = N_a \int_0^t J_a(t')[1 - \alpha(t')] \, dt' \qquad [2.114]$$

The resolution of equation [2.111], especially when the nucleation and growth rates $J(t)$ and $\dot{R}(t)$ are time-dependent, is a complex mathematical problem. Usually, two simplified cases are considered, namely the *instantaneous* and *progressive* nucleations.

2.5.1.1. *Instantaneous nucleation*

Instantaneous nucleation corresponds to the following assumptions:

1) the N_a available nucleation sites nucleate at once at $t = 0$;

2) then they grow at a constant rate $\dot{R}_0$.

Condition 1 is satisfied when supersaturation is large enough for the nucleation sites to instantly nucleate. Formally, this condition translates into:

$$J_a(t) = \delta_D(t) \qquad [2.115]$$

where $\delta_D(t)$ is the Dirac function. This implies that:

$$\int_0^t J_a(t') \left(\int_{t'}^t \dot{R}(t'') \, dt'' \right)^3 dt' = \left(\int_0^t \dot{R}(t'') \, dt'' \right)^3 \qquad [2.116]$$

Considering condition 2, the integral in equation [2.116] is easily computed, which yields:

$$\alpha(t) = 1 - \exp\left(-\left(\frac{t}{\tau_{inst}} \right)^3 \right) \qquad [2.117]$$

where

$$\tau_{inst} = \left(\frac{3}{4\pi N_a} \right)^{1/3} \frac{1}{\dot{R}_0} \qquad [2.118]$$

is the time constant of instantaneous nucleation.

As expected in this case, equation [2.115] leads by application of equation [2.114] to:

$$N(t) = N_a \qquad [2.119]$$

2.5.1.2. Progressive nucleation

Progressive nucleation corresponds to the following assumptions:

1) the nucleation rate is constant and corresponds to its stationary value $(J_a(t) = J_{a,s})$;

2) as in the previous case, the growth rate is supposed to be constant $(\dot{R}(t) = \dot{R}_0)$.

Similarly to the previous case, equation [2.111] is easily transformed into[11]

$$\alpha(t) = 1 - \exp\left(-\left(\frac{t}{\tau_{prog}}\right)^4\right) \qquad [2.120]$$

where

$$\tau_{prog} = \left(\frac{3}{\pi N_a J_{a,s} \dot{R}_0^3}\right)^{1/4} \qquad [2.121]$$

is the time constant of progressive nucleation.

Both regimes are compared in Figure 2.8, showing that phase transformation is delayed under the progressive nucleation regime, but is nevertheless completed earlier.

In the case of progressive nucleation, the final concentration of supernuclei is given by:

$$N_{max,prog} = N_a J_{a,s} \tau_{prog} \int_0^\infty \exp\left(-u^4\right) du \propto \left(\frac{N_a J_{a,s}}{\dot{R}_0}\right)^{3/4} \qquad [2.122]$$

---

11 For this purpose, the integral is computed with the following variable change: $u = t - t'$.

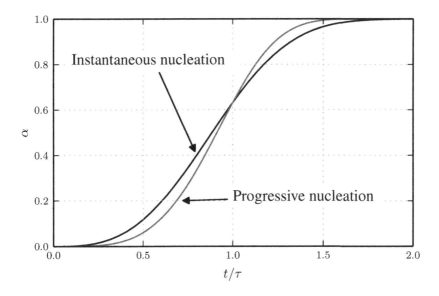

Figure 2.8. *Time evolution of the transformed volume fraction α with time under instantaneous and progressive nucleation regimes*

2.5.2. *Exclusion zones*

It is often necessary to take into account the formation, around growing supernuclei, of an area where nucleation is inhibited, for instance because monomers are depleted (nucleation and growth are then fed by mass transport of the monomers from the bulk of the old phase). The extension of the Johnson-Mehl-Avrami-Kolmogorov equation (JMAK) formalism to this case is straightforward. Indeed, it suffices to introduce the volumic fraction of the old phase where nucleation is still active, which is given by a similar equation as equation [2.111]:

$$1 - \alpha_{ex}(t) = \exp\left[-\frac{4\pi}{3} N_a \int_0^t J_a(t') \left(\int_{t'}^t \dot{R}_{ex}(t'') \, dt'' \right)^3 dt' \right] \qquad [2.123]$$

where R_{ex} is now the radius of the spherical exclusion zones surrounding supernuclei.

Consequently, new supernuclei can only form in active areas of the old phase, i.e. in the volume $V[1 - \alpha_{ex}(t)]$. Therefore, equation [2.42] leads to:

$$N(t) = N_a \int_0^t J_a(t')[1 - \alpha_{ex}(t')]\, dt' \qquad [2.124]$$

The same cases of instantaneous and progressive nucleation can be described. We shall use this case to illustrate the situation when exclusions zone grow as follows[12]:

$$R_{ex}(t) = (v_{ex}t)^{1/2} \qquad [2.125]$$

which is another growth regime often considered.

2.5.2.1. Instantaneous nucleation

A similar treatment as in section 2.5.1 leads to:

$$\alpha_{ex}(t) = 1 - \exp\left(-\left(\frac{t}{\tau_{ex,inst}}\right)^{3/2}\right) \qquad [2.126]$$

$$\tau_{ex,inst} = \left(\frac{3}{4\pi N_a}\right)^{2/3} \frac{1}{v_{ex}^{3/2}} \qquad [2.127]$$

Again, as a direct consequence of the Dirac shape of $J_a(t)$, it is easy to demonstrate that equation [2.124] passes into $N(t) = N_a$.

2.5.2.2. Progressive nucleation

This case is also treated similarly as discussed in section 2.5.1 by the evaluation of the following integral:

$$N_a \int_0^t J_{a,s} v_{ex}^{3/2} [t^{1/2} - t'^{1/2}]^3\, dt' \qquad [2.128]$$

---

12 In section 2.5.1, the growth was described by: $R(t) = \dot{R}_0 t$, where $\dot{R}_0$ was a constant.

The variable change $u = t^{1/2} - t'^{1/2}$ easily leads to:

$$\alpha_{ex}(t) = 1 - \exp\left(-\left(\frac{t}{\tau_{ex,prog}}\right)^{5/2}\right) \qquad [2.129]$$

$$\tau_{ex,prog} = \left(\frac{5}{2} \frac{3}{4\pi N_a J_{a,s} v_{ex}^{3/2}}\right)^{2/5} \qquad [2.130]$$

Accordingly, equation [2.124] yields:

$$N(t) = N_a J_{a,s} \tau_{ex,prog} \int_0^{t/\tau_{ex,prog}} \exp(-u^5) du \qquad [2.131]$$

And the final concentration of supernuclei reads:

$$N_{ex,max,prog} = N_a J_{a,s} \tau_{ex,prog} \int_0^\infty \exp(-u^5)\, du \propto \left(\frac{N_a J_{a,s}}{v_{ex}}\right)^{3/5} \qquad [2.132]$$

2.6. Conclusions

There are several ways to describe the emergence of a new phase from an old phase. In this process, the central mechanism is certainly the nucleation of extremely small domains of the new phase. In this chapter, we have focused on the granular description of nucleation in which the question reduces to the determination of the energetic cost associated with the formation of the nucleus. We have restricted ourselves to the so-called classical nucleation theory to address this question.

Then, growth mechanisms have been discussed. The attachment of monomers to clusters of the new phase has been examined. This process is also relevant to describe the growth of macroscopic domains of the new phase.

An alternative mode for the progression of the new phase is the aggregation and coalescence of clusters. Aggregation is defined here as the process by which two clusters collide into one larger domain. This process may be described using the same formalism as nucleation (nucleation is the aggregation of monomers with larger clusters). Coalescence is the process by which supernuclei grow enough to come into contact. The evolution of transformed   volume   fraction   can   be   predicted   using   the Johnson-Mehl-Avrami-Kolmogorov equation (JMAK) formalism.

Finally, we now possess suitable formalisms to describe all stages of phase transformation – or more exactly the formation of a condensed phase. In Chapter 3, this knowledge is applied to more specific cases of metal electroplating and the synthesis of NPs.

3

The Precipitation of Metals: Thin Film Electroplating and Nanoparticle Synthesis

The concepts discussed in Chapter 2 are developed further in specific cases of metal precipitation: electroplating and NP synthesis. In both cases, a chemical (or electrochemical) reaction supplies the monomers involved in the nucleation process. This reaction can be introduced in the mathematical description as a generation term $K_1(t)$ in equation [2.11] or equation [2.106]. Beyond this common characteristic, these two processes are very different.

3.1. Principles of metal electroplating

Conceptually, metal electroplating is a rather simple process: the conductive substrate to be coated – the working electrode (WE) – is immersed in an electrolyte containing Me^{z+} ions of the metal Me to be deposited (Figure 3.1). A second electrode – the counter electrode (CE) – is also placed in the solution. Using an external generator, an electric potential is applied between the two electrodes to force the following reaction on the WE:

$$Me^{z+} + ze^- \longrightarrow Me \tag{3.1}$$

As this reaction is a reduction, the WE acts as a cathode. Consequently, an oxidation reaction needs to occur at the CE, which is thus an anode. Often, this oxidation reaction is the reverse of the reduction [3.1]. In this case, the anode is made up of metal Me, and dissolves into Me^{z+}, replenishing the electrolyte with the ions consumed at the cathode. Such an anode is referred to as a *soluble anode*. Alternative anodic systems are also possible. For instance, if *inert* anodes are used (such as Pt), the anodic reaction may be the oxidation of the solvent, which is usually water:

$$H_2O \longrightarrow 2\,H^+ + 2\,e^- + \frac{1}{2}\,O_2 \qquad\qquad [3.2]$$

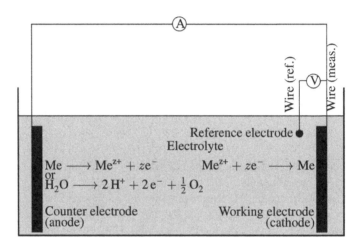

Figure 3.1. *Schematic representation of an electrochemical cell for metal electroplating*

Generally, electrolytic deposition processes are designed to minimize the contribution of the anodic reaction in the system's response. Therefore, *all electrochemical phenomena of interest regarding the solidification occur at the surface of the WE (cathode)*. In the following sections, they are introduced in the light of the concepts discussed in Chapter 2.

3.1.1. Thermodynamic considerations

3.1.1.1. The Nernst equation

Following the reasoning in Chapter 2, it is first necessary to examine the system under equilibrium. In the case of metal electroplating, this equilibrium implies a balance of equation [3.1] so no net reaction is observed at the surface of the WE:

$$Me^{z+} + ze^- \rightleftharpoons Me \qquad [3.3]$$

Each of the species involved in this equation possesses its own chemical potential as introduced in section 2.1.1. In the case of charged species (such as Me^{z+} ions or electrons), an additional contribution must be taken into account, which is the local electric field [TRE 93]. This leads to the definition of the *electrochemical potential*:

$$\tilde{\mu}_{\alpha,i} = \mu_{\alpha,i} + z_i F \Phi_\alpha \qquad [3.4]$$

where $\mu_{\alpha,i}$ is the chemical potential defined in equation [2.2], z_i is the number of charges borne by the considered species, $F = 96{,}485 \ \mathrm{Cmol^{-1}}$ is Faraday's constant[1] and Φ_α is the *internal* electric potential (or Galvani potential) of phase α. Therefore, an alternative formulation of the equilibrium relation [3.3] is:

$$\tilde{\mu}_{Sol,Me^{z+}} + z\tilde{\mu}_{El,e^-} = \tilde{\mu}_{El,Me} \qquad [3.5]$$

where the subscripts El and Sol refer to the electrode and the solution, respectively. Considering the definition of the electrochemical potentials equation [3.4], this equation may be rewritten into:

$$\left(\Phi_{El} - \Phi_{Sol}\right)_{eq} = \Delta\Phi_{eq} = \frac{1}{zF}\left(\mu_{Sol,Me^{z+}} + z\mu_{El,e^-} - \mu_{El,Me}\right) \qquad [3.6]$$

where the subscript eq indicates the equilibrium state. This relation can be further developed considering that the chemical potentials of dissolved species can be expressed as [GUG 85]:

$$\mu_{\alpha,i} = \mu_{\alpha,i}^0 + RT \ln a_{\alpha,i} \qquad [3.7]$$

---

1 When dealing with chemical solutions, the usage is to consider *molar* quantities, which will be done from now on.

where $R = 8.31\,\mathrm{J\,mol^{-1}K^{-1}}$ is the gas constant and $a_{\alpha,i}$ is the *activity* of solute i in phase α. Therefore, equation [3.6] can be transformed into:

$$\Delta\Phi_{eq} = \Delta\Phi^0 + \tfrac{RT}{zF}\ln a_{Sol,Me^{z+}}$$

$$\Delta\Phi^0 = (\Phi_{El} - \Phi_{Sol})^0 = \tfrac{1}{zF}\left(\mu^0_{Sol,Me^{z+}} + z\mu_{El,e^-} - \mu_{El,Me}\right) \qquad [3.8]$$

The Galvani potential of a phase is not accessible to experimental measurements. Fortunately, according to equation [3.6], it is the difference between the internal potentials of the electrode and the solution that needs to be measured. This can be done using a potentiometric setup composed of the electrode of interest (WE), and a second electrode with a known potential called reference electrode (RE) (Figure 3.1). The latter usually contains a solution of controlled composition[2]. Therefore, the measurement chain can be schematically represented by:

Wire (ref.) | RE | Ref. solution Electrolyte | WE | Wire (meas.)

The potential difference E between the two wires results from the difference between Φ_{El} and Φ_{Sol}, with additional contributions of the other interfaces, including the porous junction between the reference and electrolyte solutions. As all the latter contributions are constant, we find that:

$$E_{eq} = \Delta\Phi_{eq} + \text{constant} \qquad [3.9]$$

The constant terms in equations [3.8] and [3.9] are gathered in the so-called *standard potential* E^0. Therefore, the equilibrium potential of the metal electrode reads:

$$E_{eq} = E^0 + \frac{RT}{zF}\ln a_{Sol,Me^{z+}} \qquad [3.10]$$

which is known as the *Nernst equation*. By convention, the potentials E_{eq} and E^0 are usually expressed with respect to the standard hydrogen electrode.

---

2 A typical reference is the saturated calomel electrode (SCE) whose potential is $+0.248\,\mathrm{V}$ versus the standard hydrogen electrode (SHE).

For practical purposes, the activity of Me^{z+} in solution is approximated by its concentration (this approximation is more accurate for dilute solutions). Therefore, the Nernst equation becomes:

$$E_{eq} \simeq E^0 + \frac{RT}{zF} \ln [Me^{z+}] \qquad [3.11]$$

3.1.1.2. Conventional model for metal electroplating from a solution

In the most widespread approach, metal electroplating is described as follows [TRE 93]:

– the old phase is the solution containing $[Me^{z+}]_{El}$ ions (which are the monomers);

– the new phase is the metal deposited at the surface of the WE;

– the process is isothermal.

As explained in section 2.1.2, the deposition of the metal is driven by a gain in Gibbs energy, the supersaturation. In the case of electrochemical processes, the electrochemical potentials must be used to evaluate this quantity. The displacement of the equilibrium equation [3.3] toward the metal implies that:

$$\tilde{\mu}_{El,Me} < \tilde{\mu}_{Sol,Me^{z+}} + z\tilde{\mu}_{El,e^-} \qquad [3.12]$$

The supersaturation is thus given by:

$$\Delta\tilde{\mu} = \tilde{\mu}_{El,Me} - \tilde{\mu}_{Sol,Me^{z+}} + z\tilde{\mu}_{El,e^-} < 0 \qquad [3.13]$$

Using equation [3.4], this leads to:

$$\Delta\tilde{\mu} = \mu_{El,Me} - \left(\mu_{Sol,Me^{z+}} + z\mu_{El,e^-} + zF\Delta\Phi\right) \qquad [3.14]$$

where $\Delta\Phi = (\Phi_{El} - \Phi_{Sol})$ is the out-of-equilibrium potential difference between the electrode and the solution. Using now equations [3.7] and [3.8], the equation is written as:

$$\begin{aligned} \Delta\tilde{\mu} &= zF\Delta\Phi - \left(zF\Delta\Phi^0 + RT\ln[Me^{z+}]\right) \\ &= zF(\Delta\Phi - \Delta\Phi_{eq}) \end{aligned} \qquad [3.15]$$

Owing to equation [3.9], we find that to displace equation [3.3] toward metal deposition, it is needed to apply to the WE a potential E below its equilibrium potential E_{eq}. The supersaturation in the electroplating process is simply given by the difference:

$$\Delta\tilde{\mu} = zF(E - E_{eq}) = zF\eta < 0 \qquad [3.16]$$

where η is called overpotential and is readily accessible to or controllable by the experimenter.

Upon application of a potential $E \neq E_{eq}$ to the WE, the local concentration $[\text{Me}^{z+}]_{\text{El}}$ of metal ions in the vicinity of its surface evolves to satisfy:

$$\ln [\text{Me}^{z+}]_{\text{El}} \rightarrow \frac{zF}{RT}(E - E^0) = \ln [\text{Me}^{z+}] + \frac{zF}{RT}\eta \qquad [3.17]$$

As for deposition it is needed that $\eta < 0$, this means that (as expected) *the local concentration of metal ions near the cathode is lower than in the bulk of the solution.*

3.1.1.3. A less conventional model for metal electroplating from a solution

As already suggested in section 2.3.1, a more comprehensive description would include the consideration of adatoms (Me_{ad}) as intermediate species in the process of electrocrystallization. These adatoms are produced by the discharge of Me^{z+} ions and should diffuse on the surface of the electrode toward suitable crystallographic sites of the new phase (either by participating to the growth of supernuclei or to nucleation). *In this case, the old phase is composed of a 2D liquid formed by these adatoms.* As a result, equation [3.3] should be developed into:

$$\text{Me}^{z+} + ze^- \underset{\rightleftharpoons}{\overset{1}{\rightleftharpoons}} \text{Me}_{\text{ad}} \overset{2}{\rightleftharpoons} \text{Me} \qquad [3.18]$$

In this case, the monomers are produced by the reaction [3.18]–1, which contributes to the term $K_1(t)$ in equation [2.106] (see section 2.4.1). The same reasoning as in section 3.1.1.2 may be applied replacing the activity of the adatoms by their surface concentration $[\text{Me}_{\text{ad}}]$:

$$\mu_{El,Me_{ad}} = \mu^0_{El,Me_{ad}} + RT \ln [\text{Me}_{\text{ad}}] \qquad [3.19]$$

This leads to the following expression for the supersaturation:

$$\Delta\tilde{\mu} = RT \ln \frac{[Me_{ad}]_{eq}}{[Me_{ad}]}$$ [3.20]

The relation between the supersaturation and the overpotential is thus:

$$\Delta\tilde{\mu} = zF\eta - RT \ln \frac{[Me^{z+}]_{El}}{[Me^{z+}]}$$ [3.21]

where, as before, $[Me^{z+}]_{El}$ denotes the local concentration of metal ions near the electrode surface, and $[Me^{z+}]$ is the concentration of metal ions in the bulk solution. This significantly complicates the description of the system, as now $[Me^{z+}]_{El}$, which may vary over time or upon location, allows for temporal or local variations of the supersaturation (even if the overpotential is constant).

However, this has been a necessary evil in some cases where the diffusion of adatoms was claimed to be the limiting step for growth kinetics [BUD 00, FLE 66]. Therefore, we cannot exclude that this phenomenon may also contribute in the nucleation kinetics [MIL 74b].

3.1.2. Kinetics of electroplating and Faraday's Law

In addition to these phenomena, electrolytic deposition of metals may be kinetically controlled by [TRE 93]:

– mass transport of Me^{z+} ions from the solution[3];

– desolvation of these ions as they approach the electrode surface;

– the charge transfer reaction itself (equation [3.1]);

– and possibly the diffusion of adatoms (formed in the discharge step) toward crystalline sites of the metal.

These aspects will be discussed further in the next chapter. In the present discussion, we shall restrict ourselves to the practical consequences of

---

3 In Chapter 2, this effect has nonetheless been taken into account implicitly in section 2.3.1 through c_{ad}, the surface concentration of adsorbed monomers and explicitly in section 2.5.2 as limited mass transport may be the cause of the formation of exclusion zones around supernuclei.

deposition kinetics on the electrochemical response of the system. Indeed, according to equation [3.1], there exists a direct relationship between the mass balance of the reaction and the electrical charge Q transferred during electrolysis:

$$m_{Me} = \frac{Q}{zF} M_{Me} \qquad [3.22]$$

where m_{Me} is the deposited mass and M_{Me} is the molar mass of the metal. This equation is known as Faraday's law. The time derivative of equation [3.22] yields the deposition rate

$$\frac{dm_{Me}}{dt} = \frac{M_{Me}}{zF} \frac{Q}{dt} = \frac{M_{Me}}{zF} j \qquad [3.23]$$

where j is the *current density* (current per unit surface of WE) passed during electrolysis.

During electroplating, the deposition rate is directly proportional to the current density j. Therefore, local variations of j correspond to similar variations in the deposition rate.

3.1.3. *Electrochemical response of nucleation phenomena*

Whatever the considered mechanism, nucleation influences the electrochemical response of the system. In other words, the occurrence of nucleation at the surface of the electrode modifies its electrochemical behavior. This phenomenon can be captured in quite simple experiments such as cyclic voltammetry or chronoamperometry.

From the discussions in sections 3.1.1.2 and 3.1.1.3, the establishment of the supersaturation needed to drive nucleation corresponds to the imposition of an overpotential to the WE. Moreover, it has been shown in section 2.2.6.1 that even if such a supersaturation exists, nucleation may not be observed experimentally because it is kinetically inactive (i.e. its time scale, measured by τ as defined in equation [2.86], exceeds the characteristic duration of the experiment). The absence of nucleation at the electrode surface has a simple practical consequence: *no current flows through the electrochemical cell.*

Therefore, in cyclic voltammetry, upon sweeping into the cathodic domain (with respect to the equilibrium potential of Me^{z+}/Me*), a current may not be immediately measured because the metal has not nucleated yet.*

A corollary of this argument is that nucleation kinetics could be quantitatively studied by recording electrochemical transients. Indeed, upon application of a fixed overpotential, the current response contains the contribution of nucleation. Several theoretical descriptions have been proposed to extract this information from current (or potential) transients. Many of them take into account the diffusion of metal ions from the bulk solution, as this is the main cause fixing the local value of $[Me^{z+}]_{El}$. A seminal model, proposed by Sharifker and Hills, considers the growth of exclusion zones around supernuclei (see section 2.5.2), but *in the solution*, and differentiates the instantaneous and progressive regimes introduced in section 2.5.1 [SCH 83]. Several refinements have been proposed subsequently, as reviewed by Hyde and Compton [HYD 03]. More recently, models based on atomistic description of the clusters have also been proposed [MIL 74b, MIL 74a]. They will not be discussed further here.

In Chapter 5, we will examine how the electrochemical response of nucleation can guide the development of more efficient processes for the electrolytic deposition of thin films.

3.2. Chemical synthesis of metallic nanoparticles

Among the available techniques for the synthesis of metallic NPs, chemical processes are certainly the most versatile. They offer almost infinite possibilities to design the composition, shape and size of the nano-objects [SAU 10]. Also, many of these processes can be performed at or near room temperature under mild conditions. Most of them proceed by the reduction of metal ions, or by the decomposition of precursors [COR 13]. Often, these reactions are carried out in aqueous or organic solutions; because of their wider electrochemical windows, however, organic solvents give access to a larger range of metals.

3.2.1. Mechanism of formation: the Lamer model

In the chemical processes of interest in this chapter, nucleation and growth mechanisms play a central role, for instance to provide monodisperse suspensions of NPs. Indeed, following the discussion in section 2.5.1.1, this can be done when nucleation and growth phases are well separated. This can be achieved in chemical processes if:

1) the monomers are produced by some chemical reaction from some *precursor*;

2) the system is ideally closed, i.e. there is no supply of precursor into the system.

In this configuration, it is possible to obtain a specific transient nucleation regime, which was not explicitly described in Chapter 2. In this particular regime, monomer concentration (thus supersaturation) increases to reach some critical value at which nucleation starts. If the precursor is not too concentrated, soon enough the concentration of monomers drops due to the limited supply of monomers. The system enters a phase where supersaturation is not sufficient to allow for further nucleation, but where growth of the supernuclei is still active. This mechanism, discussed in [LAM 50], thus corresponds to a situation of instantaneous nucleation (Figure 3.2).

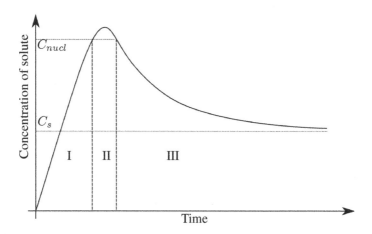

Figure 3.2. *Schematic representation of the Lamer mechanism of instantaneous nucleation of monodisperse NPs. I: pre-nucleation, II: nucleation and III: growth (C_s is the saturation concentration of the solute and C_{nucl} is the threshold concentration for nucleation). Adapted from [LAM 50]*

3.2.2. *Mechanisms of stabilization*

As mentioned in section 2.4.2, the aggregation of NPs is driven by the action of Van der Waals attractive forces. A first mechanism that counteracts this attraction is electrostatic repulsion. In conventional electrolytes, the distribution of ions is usually described by the concept of electrical double layer (EDL). This structure accommodates the mismatch in internal potential between the solid and the liquid (Figure 3.3(a)). Depending on the relative values of both potentials, cations or anions preferentially stick to the surface, forming the *internal layer* (also called Helmoltz layer). This layer accommodates most of the potential mismatch. The rest causes a more diffuse layer close to the surface, referred to as the *Gouy–Chapman layer*. This formalism can be transposed in the case of NPs. As a result of the formation of an EDL around them, NPs experience a repulsive electrostatic force in addition to the attractive Van der Waals interaction (Figure 3.3(b)). This superposition of antagonist potentials is the basis of the so-called Derjaguin–Landau–Verwey–Overbeek (DLVO) theory and explains why, for instance, aqueous suspensions are less stable at higher ionic strengths when EDL forces are screened.

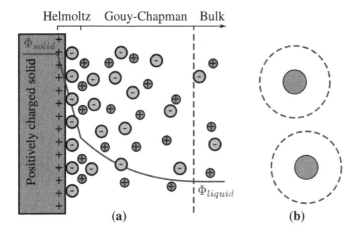

(a)   (b)

Figure 3.3. *a) Schematic representation of the EDL in conventional electrolytes and b) corresponding modified area around metallic NPs responsible for electrostatic repulsion in the DLVO theory. For a color version of this figure, see www.iste.co.uk/haumesser/metals.zip*

However, electrostatic repulsion may not be strong enough to durably maintain NPs in suspension. In a more efficient strategy, steric effects are employed to stabilize NPs. For this purpose, molecules such as ligands, polymers or surfactants are adsorbed into NPs, leaving no free metallic surface for agglomeration [PAC 08]. Therefore, suitable compounds need to efficiently coordinate with the metal, but also be solvated by the solvent. Otherwise, "solvo-phobic" interactions may lead to sticking and agglomeration of the coordinated NPs. These ligands can also influence the size and shape of the NPs, adding degrees of freedom in the process [SAU 10]. In addition, they may bear functional groups that modify the reactivity of the NPs to fit with the desired application [GOE 10]. The affinity of these ligands with the NPs may also be varied to tune the controlled aggregation discussed in section 2.4.2 [WAN 14].

However, there are situations in which the surface of the metallic NPs need not be modified or masked by such stabilizing agents. In many applications, NPs are used for their catalytic properties. Obvious examples are chemical reactions of industrial or environmental interest [FRE 11]. In nano-sciences, these catalytic properties are also used to grow specific nano-objects, such as nanotubes [DIJ 10] or nanowires [REN 09]. In special cases, these NPs can be used as precursors for the deposition of conformal films, such as the seed layer discussed in Chapter 5 [HAM 12]. In this case, the metal must be as pure as possible to limit in-film contamination [BAR 08]. Therefore, for all these applications, ligand-free NPs are needed, sometimes even in the liquid phase. A simple and low-cost process capable to produce such accessible NPs is discussed in detail in Chapter 6.

3.3. Conclusions

Electroplating and NP synthesis processes are two different cases of metal precipitation. Electroplating involves heterogeneous nucleation, whereas NP synthesis is a homogeneous phenomenon preferentially occurring through instantaneous nucleation. Electroplating aims at forming thin films with early coalescence, whereas aggregation must be avoided in NP synthesis. In both cases, though, the nucleation step is important in determining the final characteristics of the objects (either thin film or NPs).

All these aspects will be developed further and exemplified in Chapters 5 and 6. Before that, we shall apply in the following chapter the knowledge of metal electroplating to the controlled (superconformal) growth of Cu films for the fabrication of advanced interconnects.

4

Copper Electroplating: from Superconformal to Extreme Fill

In Chapter 1, the technological importance of interconnect structures in integrated circuits has been introduced. To fabricate these structures, many process steps are needed. In particular, several deposition techniques for the formation of thin metallic films are involved. Each of these processes is exemplary of particular aspects of the solidification of the metal. In this chapter, we shall focus on copper electroplating as a solution to fill the hollow conducting lines or contact holes (vias). As briefly explained in Chapter 1, the void-free filling of these structures requires a specific growth regime called superconformal. We will now examine how a kinetic control of the electrodeposition of Cu may promote this particular regime in the case of damascene interconnects and TSVs.

4.1. Copper electroplating

4.1.1. *Specificities of Cu electroplating*

In the case of Cu electroplating, equation [3.3] becomes:

$$Cu^{2+} + 2e^- \rightleftharpoons Cu \qquad [4.1]$$

Therefore, all the equations in section 3.1 may be used with $z = 2$.

However, this is a simplification of the actual processes occurring during Cu electrodeposition. Indeed, it has been shown that Cu^{2+} ions do not discharge in one step as suggested by equation [4.1], but in two consecutive steps:

$$Cu^{2+} + e^- \rightleftharpoons Cu^+ \qquad [4.2]$$

$$Cu^+ + e^- \rightleftharpoons Cu \qquad [4.3]$$

Moreover, it has been shown that the limiting step is the reduction of Cu^{2+} into Cu^+ [MAT 59][1].

4.1.2. Is electroplating capable of filling cavities?

As described in section 1.1.2, the goal of the electroplating process is to fill hollow trenches or vias patterned in a silicon substrate. Hence, the question is now: is electrolytic deposition naturally capable of doing so? Based on the concepts developed in Chapter 3, mainly because Cu electroplating obeys Faraday's law, the answer is: *no, it should not.*

Indeed, several phenomena contribute to spontaneously establishing a higher current density on top of the features:

– the sharp edges of the trenches and vias concentrate electric field, and "attract" current lines, leading to locally higher current density;

– the seed layer resists electric current. Hence, there is an additional electric resistance to overcome for the current flowing to the bottom of the features;

– during deposition, Cu^{2+} ions are consumed from the electrolyte and need to be replenished by mass transport. This replenishment is slower within the features.

Depending on the geometrical characteristics of the hollow shapes (small damascene trenches or deep TSVs), the relative weight of these contributions may vary (see section 4.3.2). Nevertheless, in all cases, they lead to a significantly slower deposition at the bottom of the features. In this *subconformal* regime, cavities (also called voids) remain after electroplating (Figure 1.5).

If the natural behavior of an electroplating process is subconformal deposition, how did researchers at IBM obtain a superconformal regime (see

---

1 Cu^+ is not stable in aqueous solutions, and spontaneously disproportionates into Cu^{2+} and Cu according to:

$$2\,Cu^+ \rightleftharpoons Cu^{2+} + Cu$$

Figure 1.5)? How robust is such a regime? Can it be obtained even in deep structures such as TSVs? These questions are addressed in the following sections.

4.2. Superconformal fill of damascene trenches

Historically, Cu electroplating has been introduced as the process of choice to fill the submicron trenches and vias in the damascene sequence [AND 98]. To achieve superconformal deposition, selected additives have been incorporated into the electrolyte. In this section, we shall investigate the mechanisms of action of these chemical compounds. It will be shown that simple electrochemical experiments can bring useful information about these mechanisms. Finally, the consistency of these results with actual fill performances will be examined and discussed.

4.2.1. The electrolytes for superconformal Cu electroplating

4.2.1.1. The various types of Cu electrolytes

The electrolytes used for damascene Cu electroplating share the same base formula. They are (quite concentrated) aqueous solutions of copper sulfate, which is the source of Cu^{2+} ions. They also contain sulfuric acid, which is a supporting electrolyte. Indeed, H_3O^+ ions are the main charge carriers in the electrolyte, needed to reach high enough current densities to sustain the required deposition rates (see equation [3.23]). Initially, the *high acid* Cu electrolytes contained as much as $2\,mol \cdot L^{-1}$ sulfuric acid [NIS 07]. However, as the interconnect structures were miniaturized, thinner Cu seed layers had to be used. These more resistive films caused an increase in the so-called *terminal effect*, which corresponds to faster deposition of the metal near the wafer edge, close to the cathodic contacts [ARM 11]. In addition, the thin Cu seed layers were no longer able to withstand corrosion in the extremely acid Cu electrolytes. For these reasons, the acid concentration was drastically decreased to about $0.1\,mol \cdot L^{-1}$ in so-called *low acid* (LA) electrolytes. However, as will be shown in the following, the low acidity may impact the filling performances of the electrolytes. Therefore, electrolytes with intermediate acid concentration recently emerged, referred to as *medium acid* (MA) electrolytes [DA 05].

4.2.1.2. The organic additives

As explained in section 4.1.2, electrolytes containing only copper sulfate and sulfuric acid would form subconformal deposits if applied on damascene structures.

In order to invert the current balance between the top and the bottom of the features and obtain superconformal deposition, small amounts of additives are incorporated in the electrolytes. They belong to two antagonist families: suppressors and accelerators.

Suppressors are chemical compounds that inhibit deposition. They are usually polymers which adsorb and form an organic film on the Cu surface of the substrate [REI 01]. This adsorbate blocks the surface from the solution and prevents discharge of Cu^{2+} ions. Polyethers such as polyethylene glycol are typical suppressors for damascene electroplating [VER 05]. To adsorb on the metal, they need halogenides. For this reason, low concentrations of Cl^- ions are usually present in the electrolytes. Depending on their molecular mass, the suppressors' inhibiting power may vary. The best results are often obtained by using polymolecular mixtures.

A typical accelerator is bis(sodiumsulfopropyl)disulfide (SPS) $(NaSO_3-C_3H_6S)_2$. This compound is known to complex Cu^+, which is an important intermediate species in Cu electroplating (see section 4.1.1). Moreover, SPS and its by-products (including Cu(I) complexes) strongly adsorb on Cu [VER 05]. For this reason, this compound is expected to facilitate charge transfer.

4.2.1.3. Electrolytes

As briefly mentioned above, LA electrolytes are reputed to modify the activity of the additives and compromise filling performances of the electrolytes. In this section, our aim is not only to understand the mechanism of action of the additives, but also to demonstrate the impact of acidity on their performances. For this purpose, two electrolytes will be compared throughout this section, whose composition is given in Table 4.1. It is proposed to compare an low acid (LA) electrolyte and an medium acid (MA) electrolyte containing the same set of additives. It should be noted that the exact nature of the additives is not known (this information is confidential). However, it is highly probable that the accelerator is SPS or a derivative.

4.2.2. Suppressor and accelerator: an electrochemical study

In this section, the action of the additives will be examined using conventional electrochemical techniques.

	LA	MA
[CuSO$_4$]	0.63 mol · L^{-1}	
[H$_2$SO$_4$]	0.1 mol · L^{-1}	0.6 mol · L^{-1}
[Cl$^-$]	50 ppm	
[Suppressor]	3 mL · L^{-1}	
[Accelerator]	10 mL · L^{-1}	

Table 4.1. *Composition of the tested electrolytes for damascene electroplating*

4.2.2.1. *Electrochemical techniques*

Cyclic voltammetry is usually performed to obtain a general picture of the behavior of an electrochemical system (in our case an additive-free electrolyte, electrolyte containing one or both additives). In this technique, the potential of the WE is linearly swept back and forth, and the current is plotted as a function of the applied potential.

A typical voltammogram obtained on a Pt electrode in the additive-free MA electrolyte is depicted in Figure 4.1 (a similar behavior is observed for the LA electrolyte). The initial potential of the Pt surface is about 300 mV/SCE. From this value, the potential is swept toward negative values. During this forward cathodic sweep, no current is recorded until the potential reaches 40 mV/SCE. Below that potential, a cathodic current appears, with a first small peak around −50 mV/SCE attributed to the reduction of Cu(I) species (probably CuCl traces, as the additive-free electrolyte contains 50 ppm Cl$^-$). Then, the cathodic current steadily grows, corresponding to Cu deposition. After −400 mV/SCE, the current plateaus, which is typical of a regime where the reduction reaction is limited by mass transport of Cu^{2+} from the electrolyte to the electrode surface [TRE 93]. During the backward sweep, the current response is identical and cannot be distinguished from the forward sweep. The reaction progressively slows down and stops at about 50 mV/SCE. This corresponds to the equilibrium potential E_{eq}[2]. As the potential continues to increase, an anodic current appears, due to the anodic dissolution of the metal deposited during the cathodic sweep. As expected,

---

2 Considering the values in Table 4.1, application of equation [3.11] yields $E_{eq} \simeq 0.0$ V/SCE (with $E^0 = 0.342$ V/SHE [LYD 85]). The experimental value of 50 mV/SCE is thus in reasonable agreement with the Nernst potential.

when all the Cu is dissolved, the reaction stops and the current drops to zero[3]. After that, no additional reaction is observed during the anodic sweep.

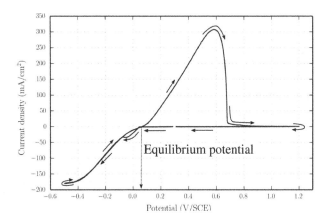

Figure 4.1. *Cyclic voltammetry response of the MA electrolyte showing Cu deposition (below equilibrium potential) and dissolution (above equilibrium potential) – sweeping rate* $10\,\mathrm{mV \cdot s^{-1}}$*, Pt RDE at* $100\,\mathrm{rpm}$

Another very simple, yet very useful technique is chronopotentiometry. It consists of imposing a current step to the system and measuring the potential response of the WE. As will be shown later, when additives are injected in the electrolyte, this potential may shift. Both the extension of this shift and the time needed for the potential to stabilize to its new value provide useful information on the behavior of the additives.

4.2.2.2. The accelerator

Let us examine how the addition of accelerator alone modifies the electrochemical response of the system.

Quite surprisingly, the cyclic voltammograms in Figure 4.2 show that the addition of the accelerator has a slight inhibiting effect when a Pt electrode

---

3 By integrating the anodic charge, it is thus possible to estimate from the voltammogram the quantity (thus the thickness) of Cu that was deposited during the cathodic sweep. This principle is used to measure the impact of the additives in the so-called cyclic voltammetry stripping experiments and monitor the composition of industrial electrolytes.

is used. In both the cathodic and anodic regions, less current flows through the cell when the accelerator is present. This can be attributed to the strong adsorption of this additive (probably by its −SH end) on Pt.

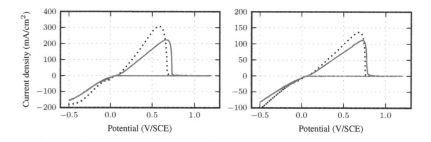

Figure 4.2. *Cyclic voltammetry response of accelerator (solid line) in the (left) MA and (right) LA electrolytes, compared to the virgin electrolytes (dotted line) – sweeping rate* $10 \, \text{mV} \cdot \text{s}^{-1}$, *Pt RDE at* $100 \, \text{rpm}$

Similarly, the addition of the accelerator does not significantly modify the response of the system when a Cu surface is considered (Figure 4.3). In this case, a very small increase in cathodic current is observed. In other words, the accelerator does not significantly accelerate Cu deposition.

However, a noticeable feature in Figure 4.3 is the apparition of a small cathodic peak at low overpotential when the accelerator is added. As already discussed in the case of Pt (see Figure 4.1), this corresponds to the reduction of adsorbed Cu(I) species. The presence of this peak is evidence for the enhanced formation of such species by the accelerator.

Finally, and as expected from the results in cyclic voltammetry, the accelerator does not significantly modify the chronopotentiometric response of the system (Figure 4.7, blue dotted curves). Upon addition of the accelerator, the electrode potential is progressively and weakly shifted toward less cathodic values, showing that the reaction is slightly favored. It should be noted that the stabilization of the potential takes several tens of seconds.

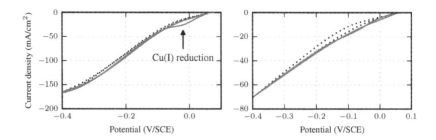

Figure 4.3. Cyclic voltammetry response of accelerator (solid line) in the (left) MA and (right) LA electrolytes, compared to the virgin electrolytes (dotted line) – sweeping rate $10\,\mathrm{mVs{-}1}$, Pt RDE with Cu predeposit at $100\,\mathrm{rpm}$

To conclude, in spite of its name, the accelerator is not really accelerating deposition as compared to the additive-free electrolyte. Beyond that quite confusing conclusion, it should be noted, however, that this additive does actually modify the surface chemistry of Cu, with the enhanced formation of Cu(I) species, which are known intermediates in the reduction of Cu^{2+} ions into Cu metal (see section 4.1.1).

4.2.2.3. The suppressor

Certainly, the suppressor has a much more spectacular effect on the electrochemical response of the system. From the cyclic voltammograms in Figure 4.4, the inhibiting action of the suppressor is obvious. Globally, much less cathodic and anodic currents flow through the cell, showing that both Cu deposition and dissolution are inhibited. These reactions are even extinguished at low overpotentials: the current is zero in a significant potential range around the equilibrium potential of Cu.

The existence of a cathodic overpotential to trigger Cu deposition is also visible when a Cu electrode is used (Figure 4.4). Therefore, this additive behaves similarly on both Pt and Cu surfaces. Interestingly enough, this overpotential can be interpreted as an excess supersaturation to initiate Cu nucleation on the electrode when the suppressor is adsorbed (see section 3.1.3). Indeed, this adsorbate would block nucleation sites, similarly to the situation described in section 5.1.3.

The inhibiting action of the suppressor can be more precisely quantified using chronopotentiometry. Indeed, the addition of suppressor in the system causes a spectacular shift in the deposition potential, as can be seen in Figure 4.7 (green dashed curve). After the addition of the suppressor, the potential immediately drops to about $-275\,\mathrm{mV/SCE}$. To sustain the imposed deposition rate, an overpotential of $-250\,\mathrm{mV}$ is thus needed when the suppressor is present (without additive, the potential stays at about $-25\,\mathrm{mV/SCE}$). *This overpotential is thus a direct measure of the inhibiting strength of the suppressor.* Moreover, the fast response of the potential to the addition of suppressor shows that this additive quickly adsorbs to the metal surface.

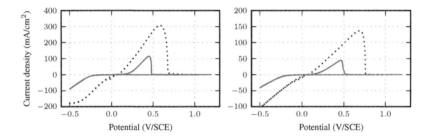

Figure 4.4. *Cyclic voltammetry response of suppressor (solid line) in the (left) MA and (right) LA electrolytes, compared to the virgin electrolytes (dotted line) – sweeping rate* $10\,\mathrm{mV\cdot s^{-1}}$*, Pt RDE at* $100\,\mathrm{rpm}$

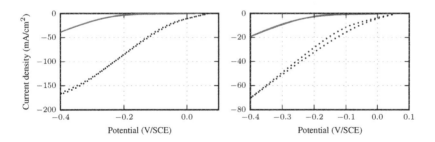

Figure 4.5. *Cyclic voltammetry response of suppressor (solid line) in the (left) MA and (right) LA electrolytes, compared to the virgin electrolytes (dotted line) – sweeping rate* $10\,\mathrm{mV\cdot s^{-1}}$*, Pt RDE with Cu predeposit at* $100\,\mathrm{rpm}$

As a conclusion, the suppressor is a strong inhibitor of both Cu deposition and dissolution, probably by the formation of an organic film on the metal surface, which blocks charge transfer between Cu^{2+} and Cu. Chronopotentiometric measurements show that this film is quickly formed. This technique also provides a quantification of this inhibition.

4.2.2.4. The combined additives

If both additives modify the electrode surface and thus influence the deposition reaction, no electrochemical evidence of a mechanism possibly leading to superconformal fill of hollow shapes was found when adding these compounds separately. Therefore, let us examine what happens when a mixture of both additives is used.

Looking first at cathodic sweeps on a Cu electrode, it can be seen that, as expected, an intermediate current is measured over the whole potential range (Figure 4.6). The deposition is less inhibited than when only the suppressor is added, but still less active than in the additive-free or accelerator-containing electrolyte. However, and in contrast with the previous situations, the forward and backward sweeps do not superimpose. During the forward sweep, only a small current is measured, similar to the response with the suppressor only. During backward sweep, a larger current is observed, closer to the response with the accelerator only.

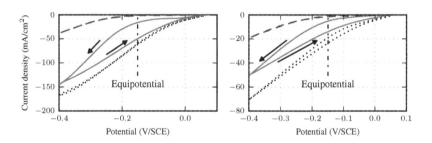

Figure 4.6. *Cyclic voltammetry response of combined accelerator and suppressor (solid line) in the (left) MA and (right) LA electrolytes, compared to the virgin electrolytes (dotted line) and the electrolyte containing only the suppressor (dashed line) – sweeping rate $10\,\mathrm{mV}\cdot\mathrm{s}^{-1}$, Pt RDE with Cu predeposit at $100\,\mathrm{rpm}$*

As a result, a *hysteretic loop* forms during cycling in the cathodic region when both additives are present. This is of paramount importance to explain superconformal filling: *the existence of such a hysteresis is necessary to allow for the coexistence, on the same (equipotential) surface, of regions with slow and fast deposition.* For instance, for a Cu electrode at $-150\,\mathrm{mV/SCE}$ in the MA electrolyte, current density can locally vary from a few $\mathrm{mA \cdot cm^{-2}}$ to $-70\,\mathrm{mA \cdot cm^{-2}}$. Obviously, this behavior is related to the nature of the adsorbed additives. Initially, the suppressor certainly dominates, accounting for the small initial current. Over time, and as potential is swept to more cathodic values, the accelerator progressively replaces the suppressor on the surface, allowing for a faster deposition. This effect holds even when the potential is swept back to its initial value, explaining why a higher current is measured during the backward sweep. Finally, it turns out that, when the two additives are present, the local current density (and thus deposition rate) depends on the relative local surface concentration of both compounds.

To further investigate this behavior, the chronopotentiometric response of the mixed additives is plotted in Figure 4.7 (red curve). Interestingly enough, immediately after the two additives are injected, the potential drops to a cathodic value close to the one measured when only the suppressor is present. This is evidence for the fact that initially the suppressor alone adsorbs on the surface. Soon enough though, the surface is progressively depolarized by the slower adsorption of the accelerator, which replaces the suppressor on the electrode.

From these curves, a characteristic time for the depolarization can be extracted. This treatment, illustrated in Figure 4.8, leads to the following observation: the displacement of the suppressor by the accelerator is significantly slower in the LA electrolyte. Therefore, it seems that the acidity indeed has an impact on the behavior (and performances) of the additives.

4.2.2.5. Conclusion

From this electrochemical study, important information has been gathered about the action of the two additives involved in the mechanisms of superconformal fill.

As expected, the suppressor significantly inhibits Cu deposition (and dissolution) by forming an organic adsorbate that prevents efficient discharge of Cu^{2+} ions (and Cu nucleation) on the surface. This adsorbate forms very quickly.

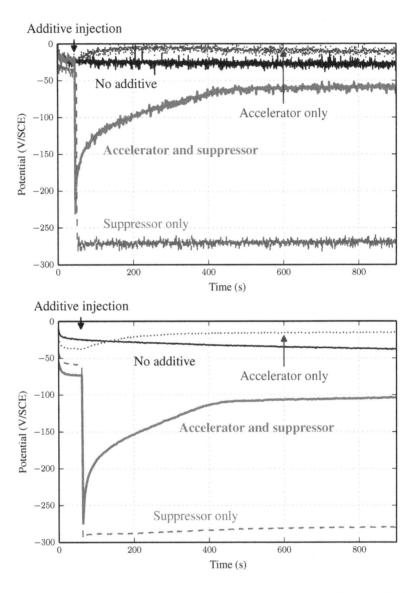

Figure 4.7. *Chronopotentiometric response of suppressor (dashed line), accelerator (dotted line) and a mixture of both (bold solid line) in the (top) MA and (bottom) LA electrolytes, compared to the virgin electrolytes (plain solid line). For a color version of this figure, see www.iste.co.uk/haumesser/metals.zip*

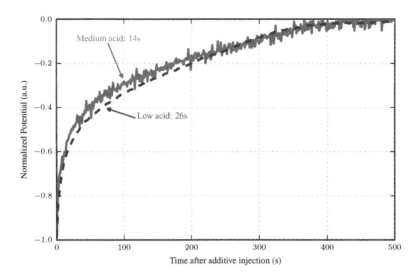

Figure 4.8. *Characteristic time for the displacement of suppressor by accelerator in the MA (solid line) and LA (dashed line) electrolytes as measured from the chronopotentiometric response of the two systems (red curves in Figure 4.7). For a color version of this figure, see www.iste.co.uk/haumesser/metals.zip*

By itself, the accelerator does not modify deposition kinetics. However, it has the ability to displace the weakly adsorbed suppressor. In this way, the deposition reaction is almost restored to its full speed. Therefore, the accelerator does not accelerate, but rather reaccelerates the inhibited deposition.

The adsorption of the accelerator is rather slow, as shown by chronopotentiometry. An important observation is that the displacement of the suppressor by the accelerator is faster in the MA electrolyte.

4.2.3. The mechanism of superconformal fill

So far, the influence of the additives on the deposition kinetics has been studied on flat surfaces. It is now time to elucidate their behavior in damascene trenches and vias.

4.2.3.1. *Can superconformal fills be quantified?*

To answer this question, let us examine the filling of typical interconnect structures corresponding to the 65 nm node. These trenches are coated with a Cu seed layer deposited by PVD, whose characteristic subconformal morphology is clear, as shown in Figure 4.9.

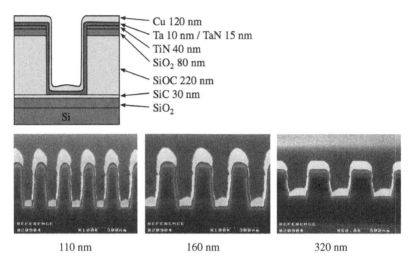

110 nm   160 nm   320 nm

Figure 4.9. *Tests structures corresponding to the 65 nm node: stack and FIB cross-sections*

The fill mechanism can be investigated by stopping the deposition at various stages. This is easily done in electroplating by interrupting the galvanostatic electrolysis at fixed charges corresponding to increasing fractions of the total charge C_{fill} needed to completely fill the trenches.

A typical set of scanning electron microscopy (SEM) cross-sections is shown in Figure 4.10. They clearly illustrate a superconformal deposition. From such images, the Cu thickness can be extracted for the bottom, the sidewall and the top of the trenches. The evolution of this thickness on these locations during plating is plotted in Figure 4.11. This graph shows the following:

– at all times during plating, the thickness on the top of the trenches (i.e. on the flat surfaces of the substrate) increases linearly at the rate of $0.4 \, \text{nm} \cdot \text{cm}^2 \cdot \text{C}^{-1}$ predicted by Faraday's law;

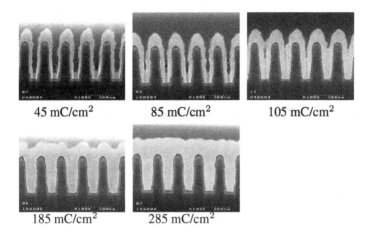

45 mC/cm²   85 mC/cm²   105 mC/cm²

185 mC/cm²   285 mC/cm²

Figure 4.10. *Partial fill experiment conducted with the LA electrolyte in* $110\,\text{nm}$ *wide trenches*

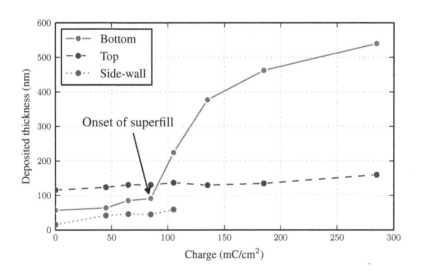

Figure 4.11. *Thickness evolution during fill with the LA electrolyte in* $110\,\text{nm}$ *wide trenches*

– initially, the same growth rate can be measured on the sidewalls and at the bottom of the trench. *During this phase, deposition is thus conformal*;

– after some time, however, the growth rate at the bottom suddenly rises to about 50 times its initial value. *This is the onset of the superconformal regime*;

– once the trench is filled, the growth rate remains significantly higher there as compared to the surrounding flat surfaces (also see Figure 4.13).

This simple experiment captures all the features of the superconformal fill, which are as follows:

– an initial conformal incubation period;

– the onset of an accelerated deposition from the bottom;

– after filling, the persistence of an accelerated growth above the trench.

A striking feature in Figure 4.11 is the abrupt onset of superconformal deposition. This affords a precise measure of the charge[4] (or time) at which the growth rate at the bottom jumps. This *critical charge* C_{crit} separates the conformal incubation ($C < C_{crit}$) from the superconformal deposition ($C > C_{crit}$). With this value, it is possible to propose a simple model to describe the filling of the trenches (Figure 4.12). During the conformal phase, the trench width varies as $l(C) = l_0 - 2v_sC$ and its depth changes according to $h(C) = h_0 - v_bC + v_tC$ with $v_b = v_s$ (conformal deposition). At the critical charge, the dimensions of the trenches are thus given by $l_{crit} = l_0 - 2v_sC_{crit}$ and $h_{crit} = h_0 - v_sC_{crit} + v_tC_{crit}$.

---

4 Strictly speaking, charge densities are reported here. They are referred to as charges for the sake of brevity.

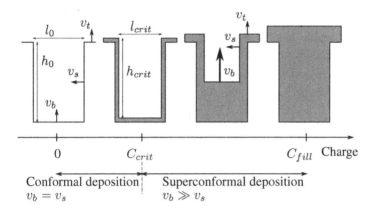

Figure 4.12. *Simple modeling of the transition between conformal and superconformal deposition regimes during Cu fill*

From these considerations, it is possible to propose a criterion that quantifies the onset of superconformal filling. It is the ratio defined by:

$$Q = 1 - \frac{l_{crit}}{l_0} \qquad [4.4]$$

This ratio represents the amount of trench width "consumed" during the conformal phase. Obviously, a Q-ratio as low as possible is desirable, as it corresponds to an early onset of the superconformal regime, which is best to ensure void-free filling.

This ratio is reported for 90 nm, 110 nm and 160 nm wide trenches filled using the MA and LA electrolytes in Table 4.2. Interestingly enough, for a given electrolyte, *the Q-ratio does not vary with trench width*. This is an important observation, as it suggests that this value can be used as an objective and quantitative criterion to evaluate the performances of a filling process, independently of the geometric characteristics of the damascene structures. From a practical standpoint, this is interesting, as large enough features may be used for accurate microscopic observations that can then be extrapolated to the small trenches of interest.

Following this line, it becomes possible to objectively and quantitatively compare the filling performances of the two electrolytes. The results in

Table 4.2 clearly show that the Q-ratio is about 27 % for the MA electrolyte, whereas it reaches 40 % in the case of the LA electrolyte. This is an interesting result, as it definitely shows that for this specific set of additives, *a lower acid concentration in the electrolyte is detrimental.*

Electrolyte	l_0 (nm)	C_{crit} (mC · cm^{-2})	Q-ratio (%)	Average (%)
MA	90	50	30	
	110	60	23	27
	160	110	29	
LA	90	85	38	
	110	85	40	40
	160	125	41	

Table 4.2. *Comparison of Q-ratio values measured in trenches of increasing width with the LA and MA electrolytes*

To summarize, the answer to the question: "can superconformal fill be quantified?" is *yes*. From quite simple – though tedious – morphological observations, the onset of superconformal deposition can be related to the amount of trench width "consumed" during the initial conformal phase. This ratio does not depend on the initial trench width, but rather on the process conditions such as electrolyte composition or applied currents.

4.2.3.2. Can superconformal fill be understood?

Even if the question of the mechanisms of superconformal deposition has attracted much interest in the scientific community since the early developments of the damascene architecture, the honest answer to this question is: *not completely.*

Indeed, experimental approaches such as those described above actually fail at capturing the detailed phenomena occurring at the growing interface during Cu electroplating. This is not only because the chemistry of this interface is complex ([VER 05] listed 19 simultaneous possible reactions), but also because *the trenches are filled in a matter of seconds, under out of equilibrium conditions* (see Figure 4.11). The acquisition of reliable experimental data with sufficient spatial and temporal resolution is thus extremely difficult, if not impossible.

For this reason, most of the models proposed to explain superconformal fill are based on the correlation of electrochemical and morphological data such as those described in the previous sections. These models all aim at answering this question: assuming that the deposition is globally inhibited by the suppressor, how is this inhibition locally lifted within the hollow shapes?

The first explanation proposed to account for the relative acceleration of deposition at the bottom of the trenches is based on a classical mechanism of *leveling by inhibiting species* [AND 98, WES 00]. In this approach, the differential growth rate mostly arises from a depletion of the inhibiting species – here the suppressor – within the features. However, this explanation was soon ruled out, because it did not account for the following experimental facts.

The depletion of the suppressor supposes that this additive is *consumed* during deposition. There are two possibilities for this: either the additive is degraded into by-products released in the electrolyte, or it is incorporated in the deposit. The suppressors are known to be hydrolyzed in acidic solutions. However, this reaction (or any other degradation reaction) was not found to be enhanced during electroplating. Similarly, the organic contamination in the electroplated Cu films is typically low, incompatible with a significant incorporation of the additives.

More importantly, this model fails to explain the persistence of an accelerated deposition once the trenches are filled. This phenomenon, already mentioned in the previous section, is the cause of a characteristic morphology of superconformal deposits, which is an area with excess metal above the filled structures (Figure 4.13 (left)). This "bump" cannot be explained by a local depletion of inhibitor. On the contrary, classical leveling mechanisms would reduce, if not eliminate such shapes. In fact, this mechanism is used to eliminate this bump, which otherwise would complicate CMP: a third additive, a *leveler*, is added to the electrolyte for that purpose (Figure 4.13 (right)).

In addition to these limitations, this simple model does not really explain the role of the accelerator in the process. The subtle interplay between the two additives is probably much better described by a more sophisticated, yet quite elegant model, called curvature-enhanced adsorbate coverage (CEAC) [MOF 05]. This model is based on a few simple hypotheses:

1) The suppressor globally inhibits deposition, and is only removed by the stronger adsorption of the accelerator, as suggested by electrochemical responses of the additives (section 4.2.2.4). As a result, the local growth rate depends on the local surface concentration (or *coverage*) of the accelerator.

2) Initially, the surface concentration of the accelerator is the same everywhere, outside or inside the hollow shapes.

3) The two additives are not incorporated in the deposit. Rather, they "float" on the moving metal surface.

4) During filling (which only takes a few seconds), the adsorbed accelerator does not move laterally on the surface (it is "pinned" to its initial location).

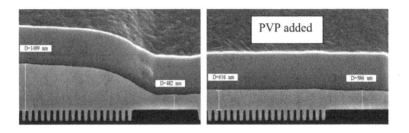

Figure 4.13. *Left: SEM micrograph of a trench array after Cu fill, showing excess metal (or* bump*) above the trenches. Right: the same Cu fill with addition of a leveler in the electrolyte. Adapted from [ZHO 07]*

These hypotheses lead to the dynamic evolution tentatively depicted in Figure 4.14. At t_0, the accelerator coverage is the same everywhere, according to hypothesis 2. Because of hypothesis 1, this causes conformal growth. During this phase, the accelerator follows the trajectories indicated in Figure 4.14 owing to hypotheses 3 and 4. As a result, it appears that the accelerator coverage is increased in the bottom corners of the trench, i.e. on the portions of convex curvature of the surface. Conversely, the accelerator is less concentrated on the concave portions of this surface, such as the trench entrance.

In the CEAC model, local variations of the accelerator concentration are explained by the curvature of the growing metal surface. In convex shapes such as hollow trenches, the accelerator concentrates during the growth. At some point, this is sufficient to fully lift the suppressor's inhibition: this is the onset of superconformal growth.

This model is successful at capturing essential features of the superconformal deposition, such as the persistent acceleration after fill. Indeed, in this model, the accumulated adsorbed accelerator remains localized above the trench. Therefore, this model readily predicts the formation of a "bump" above the filled structures (Figure 4.14(c)).

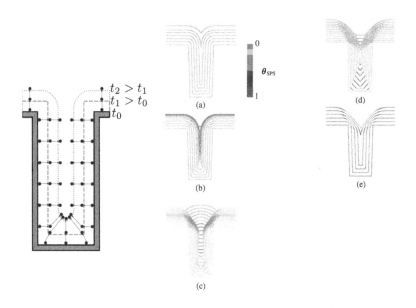

Figure 4.14. *Left: Schematic representation of the evolution of the accelerator coverage during Cu deposition. Initially, conformal deposition occurs (to dashed profile), resulting in accelerator concentration in the corners. This causes the onset of the superconformal regime (dotted profile). Right: Simulations of feature filling using catalyst-pretreated electrodes. Interface motion is displayed using colorized contour lines to reflect the local catalyst coverage. Each simulation corresponds to a different initial catalyst coverage, SPS: (a) 0.00054, (b) 0.0054, (c) 0.054, (d) 0.44 and (e) 0.88. The feature filling times corresponding to the last growth contour shown are (a) 177 s, (b) 113 s, (c) 85 s, (d) 39 s and (e) 24 s. Adapted from [MOF 05]. For a color version of this figure, see www.iste.co.uk/haumesser/metals.zip*

Along with the results of the previous section, the CEAC model leads to a sufficiently accurate description of the mechanisms of superconformal growth to efficiently guide the development of new electrolytes and the improvement of electroplating processes.

During the immersion of the sample, the suppressor immediately saturates its surface. During the initiation phase at a low current, the accelerator has enough time to partially and uniformly displace the suppressor. Interestingly enough, the electrochemical results indicate that this displacement is more extended in the MA electrolyte, and thus a larger initial coverage by the accelerator is expected.

When growth at the nominal current starts, this uniform accelerator coverage results in the conformal growth of the Cu film observed in the partial fill experiments. As a result of the CEAC mechanism, the accelerator starts to concentrate within the features.

At some point corresponding to the critical charge, this concentration is sufficient to fully restore charge exchange between Cu^{2+} and Cu, allowing the reaction to reach its full speed. If one assumes that *in all situations the same threshold accelerator coverage is needed* to do so, the extension of the conformal phase corresponds to the time needed to increase the accelerator coverage from its initial to its threshold value.

Consequently, the extension of the conformal phase only depends on the initial accelerator coverage. Indeed, according to the CEAC model, the ratio $[acc]_{thresh.}/[acc]_{init.}$ is directly related to the surface reduction at the bottom of the feature, leading to:

$$\frac{[acc]_{thresh.}}{[acc]_{init.}} = Q \qquad [4.5]$$

as defined in equation [4.4], if we neglect the contribution of the accelerator adsorbed on the sidewalls. Therefore, for given initial and threshold accelerator coverages, the onset of superconformal growth occurs at the same Q-ratio value, independently on trench width.

Moreover, this also explains the different filling performances of the LA and MA electrolytes. In the latter, the accelerator can replace the suppressor more easily, as shown by electrochemical measurements. As a result, an increased initial accelerator coverage is expected. Owing to equation [4.5], this corresponds to a smaller Q-ratio value, as observed during the partial fill experiments.

4.3. Extreme fill of TSVs

As discussed in section 1.1.3, the miniaturization of the interconnect structures is not only more and more challenging in terms of fabrication processes, but also reaches physical limits of the devices. For this reason, the alternative 3D integration is emerging, in which whole circuits are stacked and connected by deep contact holes, the TSVs.

4.3.1. *Is it easier to fill large TSVs than small damascene structures?*

TSVs are giants when compared to damascene vias (Figure 4.15). The latter have a typical width of a few tens of nanometer, and an aspect ratio (AR) of 2–3. TSVs are much wider (5 to 50 μm), which should – at first sight – make them easier to fill. However, they also are a lot deeper (as much as 100 μm deep). As a result, their AR may reach 10 or above, which becomes critical as far as metal fill is concerned.

Indeed, in such deep structures, two of the effects listed in section 4.1.2 are enhanced, namely the resistance of the seed layer and the depletion of cupric ions in the feature during electroplating. The impact of the seed layer can be mitigated as long as a thick enough metal liner can be deposited along the sidewalls and at the bottom of the TSVs, as will be shown in the following chapter. The depletion of cupric ions is directly related to the question of mass transport within the TSVs and will be discussed now.

4.3.2. *Mass transport within TSVs and depletion of Cu^{2+} ions*

During electroplating processes, bath agitation and wafer rotation provide enough convection to efficiently convey electrolyte species (Cu^{2+} and additives) to the cathode surface. Even under such conditions, the hydrodynamic stop layer that forms in the vicinity of the cathode surface extends to a few micrometer away of this surface[5]. The depth of damascene structures is small compared to this distance (Figure 4.15). Therefore, mass transport is not affected by their presence. For TSVs, the situation is completely different. Their depth is significantly larger than this hydrodynamic stop layer. To reach the bottom of a TSV, solutes have no other option than pure diffusion. Therefore, it can be expected that the bottom of the TSVs will be much less accessible (the solutes have to diffuse through the stop layer and the whole TSV) than the flat surfaces (which require only diffusion through the hydrodynamic stop layer).

The first consequence of this situation is that inside the TSVs, cupric ions are likely to be consumed faster by the reduction reaction than replenished by

---

[5] This distance may be estimated by the Levich equation for conventional RDEs: $\delta = 1.61\nu^{1/6}D^{1/3}\omega^{-1/2}$ where ν is the kinematic viscosity of the solution ($\nu = 0.01\,\text{cm}^2 \cdot \text{s}^{-1}$ for water), D is the diffusion coefficient of the electroactive species (typically $5 \times 10^{-6}\,\text{cm}^2 \cdot \text{s}^{-1}$) and ω is the rotation rate of the electrode (typically 500 rpm). These values lead to $\delta \simeq 2.3\,\mu\text{m}$.

diffusion. To obtain more quantitative insight into this matter, let us consider a very simple model.

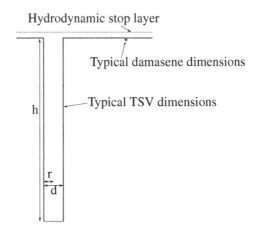

Figure 4.15. *Comparison between typical dimensions of a TSV, a damascene trench or via and the hydrodynamic stop layer*

A characteristic time for Cu^{2+} depletion can be roughly estimated using the following considerations. Let us consider a TSV of given diameter $d = 2r$ and depth h filled with an electrolyte containing a given concentration $[Cu^{2+}]$ of cupric ions and sealed to its top (i.e. there are no Cu^{2+} ions entering the structure). The electrical charge needed to consume all the Cu^{2+} entrapped in the TSV is given by:

$$Q_{depl} = 2F[Cu^{2+}]\pi r^2 h \qquad [4.6]$$

If the electroplating is assumed to occur at a uniform current density j along sidewalls and bottom, this charge corresponds to:

$$Q_{depl} = jt_{depl}(2\pi rh + \pi r^2) \approx jt_{depl}(2\pi rh) \qquad [4.7]$$

The latter approximation is accurate, as $\frac{h}{r} = 2AR$ is large. From equations [4.6] and [4.7], the time t_{depl} needed to fully deplete the TSV is thus given by:

$$t_{depl} = \frac{F[\text{Cu}^{2+}]r}{j} \qquad [4.8]$$

Similarly, the characteristic time required to replenish the cupric ions can be estimated following Fick's laws of diffusion, which define a characteristic length of diffusion l during time t_{diff} by:

$$l = \sqrt{Dt_{diff}} \qquad [4.9]$$

where D is the diffusivity of the considered species (Cu^{2+} in this case). Correspondingly, the characteristic time needed for diffusing species to reach TSV bottom is given by:

$$t_{diff} = \frac{h^2}{D} \qquad [4.10]$$

Under the approximation of equation [4.8], the time needed to deplete the TSV only depends on its diameter and is proportional to it. By contrast, the characteristic time for Cu^{2+} replenishment only depends on the TSV depth and scales with its square. Therefore, the second effect dominates the filling of deep TSVs. For this reason, from now on, we shall distinguish shallow TSVs (depth $<30\,\mu m$) from deep TSVs (depth $>30\,\mu m$).

For instance, applying equation [4.10] to a deep TSV ($h = 100\,\mu\text{m}$) and considering that the diffusion coefficient of cupric ions is $D_{Cu^{2+}} = 5 \times 10^{-6}\,\text{cm}^2 \cdot \text{s}^{-1}$ [LYD 85], the characteristic time for Cu diffusion is $t_{diff} = 20\,\text{s}$. To avoid depletion of cupric ions, the current density must not exceed a certain value, computed from equation [4.8] considering that $t_{depl} = 20\,\text{s}$. If the concentration of cupric ions is $[\text{Cu}^{2+}] = 1\,\text{mol} \cdot \text{L}^{-1}$ and considering a TSV with $d = 10\,\mu\text{m}$, this critical current density is $j = 2.4\,\text{mA} \cdot \text{cm}^{-2}$. Owing to Faraday's law (equation [3.23]), this corresponds to a Cu growth rate of $0.15\,\mu\text{m} \cdot \text{min}^{-1}$ only.

This shows that to properly fill TSVs, electroplating must necessarily be performed at low current densities. Consequently, this process can be very long, up to several hours for the deepest TSVs.

4.3.3. Electrolytes for TSV fill

These limitations imposed by mass transport within TSVs have important consequences on the implementation of an efficient fill process.

Indeed, the first idea to fill TSVs would be to use an electrolyte designed to fill damascene structures. These electrolytes are referred to as first-generation in the context of TSVs. This approach is immediately questioned by the much longer process duration to fill TSVs. Indeed, the superconformal deposition is based on the local and transient accumulation of the accelerator. If such a situation can hold during the few seconds needed to fill a damascene trench, it is certainly not the case on longer durations. With time, surface diffusion of the additive is expected to redistribute the accelerator on the surface, leading back (at best) to a conformal deposition.

To overcome this difficulty, it is possible to regularly "reset" the surface concentrations of both the suppressor and the accelerator during deposition. This can be done by regularly switching the current into an anodic polarity, partially redissolving the deposit and lifting off the adsorbed additives. Such an approach leads to a more complex current waveform, called *pulse-reverse*, as shown in Figure 4.16.

The cross-sections of a partially filled shallow TSV using a first-generation electrolyte are shown in Figure 4.17. They show that this electrolyte is successful at filling such TSVs. However, the superconformal profile of the partially filled TSV is clearly *V-shaped*. Such a profile does not correspond to the superconformal mechanism described in section 4.2.3.2: a flatter bottom would be expected. Instead, this V-shaped profile is an indication that the accelerated deposition at the bottom of the TSV is mainly caused by a depletion of the suppressor, probably due to limited diffusion of this additive with high molecular mass according to the model of leveling by inhibited species [WES 00]. Therefore, even when a pulse–reverse current waveform is used, the competitive adsorption of the accelerator does not seem to be efficient in these large holes.

This is confirmed when a similar process is applied to deep TSVs. The cross-section of a 90 µm deep TSV electroplated using a pulse-reverse current waveform (average current density below $2\,\mathrm{mA}\cdot\mathrm{cm}^{-2}$) is shown in Figure 4.18.

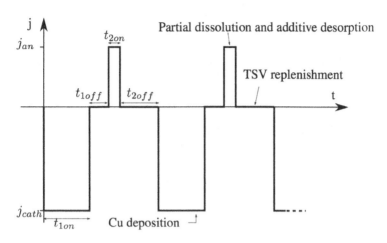

Figure 4.16. *Pulse-reverse current waveform*

Figure 4.17. *Filling a shallow TSV ($h = 17\,\mu\mathrm{m}$, $d = 3\,\mu\mathrm{m}$) with a first-generation electrolyte*

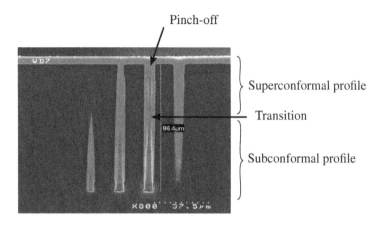

Figure 4.18. *Deep TSV ($h = 90\,\mu m$, $d = 10\,\mu m$) tentatively filled with a first-generation electrolyte*

Two cavities remain in the TSV after electroplating. The first one is small in the upper part of the via. The second one is larger, cone-shaped and placed in the lower part of the TSV. These cavities are separated by a zone where the sidewalls have merged. In this area, deposition was obviously fastest, indicating that the superconformal mechanism described above (i.e. suppressor depletion within the TSV) was active in this region. Below, the cone-shaped cavity is characteristic of a decreasing deposition rate caused by the depletion of Cu^{2+}: the deeper in the TSV, the less Cu^{2+} available, the slower the deposition.

The natural tendency would be to decrease the current density to limit Cu^{2+} depletion. However, this also increases penetration of the suppressor, reducing the superconformal effect. In fact, even at the tested current density, the superconformal growth is not sufficient to achieve defect-free fill: the small cavity in the upper part of the TSV comes from an early closure of the via entrance (referred to as *pinch-off*), probably due to the enhanced deposition near the via edges (corresponding to the first phenomenon discussed in section 4.1.2).

This study shows that the electrolytes used for superconformal fill of damascene structures are not suited for deep TSV fill. Indeed, the difference in growth rate between inhibited and accelerated locations is not sufficient to achieve defect-free fill of these deep structures, mainly because this difference only arises from depletion of the suppressor within the features, and not by accumulation of the accelerator. Therefore, new additives with enhanced activity are needed for that particular application.

Such electrolytes have recently been developed. They are referred to as third generation. As will be shown now, their behavior is different and the additives lead to a much more pronounced differential growth.

4.3.4. Electrochemical study of an electrolyte capable of extreme fill

A typical third-generation electrolyte for TSV fill – labeled Gen3 – contains two main additives: an inhibitor and an accelerator. The latter is probably similar to the one used in section 4.2 (the composition of copper electrolytes is generally confidential). Not surprisingly, cyclic voltammetry in Figure 4.19 (left) shows hardly any modification in the presence of the accelerator.

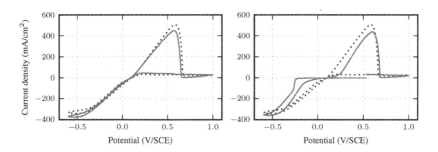

Figure 4.19. *Cyclic voltammetry response of the Gen3 electrolyte (solid line) containing only the (left) accelerator and (right) inhibitor, compared to the virgin electrolyte (dotted line) – sweeping rate* $10\,\mathrm{mV} \cdot \mathrm{s}^{-1}$*, Pt RDE at* $100\,\mathrm{rpm}$

The inhibitor in Gen3 exhibits a spectacularly different behavior in comparison with the suppressor in section 4.2. Indeed, in this case, this additive alone is sufficient to cause a hysteresis in the cathodic loop of the voltammogram (Figure 4.19 (right)). As was mentioned in section 4.2.2.4, this is a necessary condition for the establishment of a differential growth rate between (hopefully) the top and the bottom of the via. In section 4.2.2.4, this hysteresis was caused by the competitive adsorption between the suppressor and the accelerator. As the latter additive is not present here, some other mechanism must be involved.

This hysteresis is repeatable as shown in Figure 4.20, where five loops were recorded during cathodic cycling. These curves are compared to the response of an additive-free electrolyte. All five cycles exhibit the same shape. When cathodic sweep starts, no current is recorded: deposition is completely inhibited (1). At a potential of about $-300\,\mathrm{mV/SCE}$ (2), the deposition reaction starts quite abruptly. As expected, the current grows as potential is swept toward more cathodic values (3). Upon backward sweep, the cathodic current is higher, closer to its value in an additive-free situation (4). This indicates that inhibition is only partial. It is only restored once the potential returns to values close enough of the equilibrium potential (5).

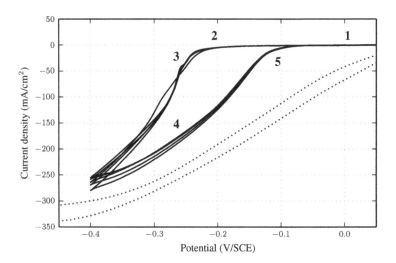

Figure 4.20. *Cyclic voltammetry response of the Gen3 electrolyte containing only the inhibitor (five cycles, solid lines) compared to the virgin electrolyte (dotted line, same as Figure 4.19) – sweeping rate* $10\,\mathrm{mV\cdot s^{-1}}$, *Pt RDE at* $100\,\mathrm{rpm}$

This behavior suggests that, conversely to the suppressors used in damascene electrolytes, this inhibitor does not reversibly adsorb on the surface. Once the inhibition is lifted, it cannot be restored as long as the cathodic potential has not returned to low enough values.

More insight into this matter can be gained by examining cathodic cycles recorded at different potential sweeping and electrode rotation rates (Figure 4.21). Comparing cycles recorded at 10 and $1\,\mathrm{mV} \cdot \mathrm{s}^{-1}$, it appears that the reduction of the sweeping rate does not modify the potential at which the inhibition is lifted (2), but shifts to more cathodic potentials its restoration (5). Globally, the hysteresis is thus narrower at low sweeping rate. This can be understood considering that under the latter condition, the inhibitor has more time to readsorb on the surface, leading to a globally more reversible response. Thus, it seems that the adsorption of this inhibitor is not immediate.

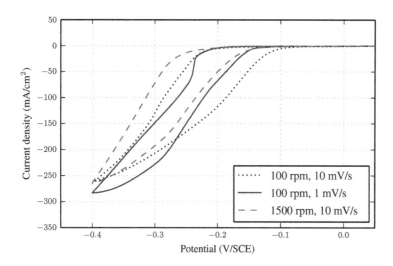

Figure 4.21. *Cyclic voltammetry response of the Gen3 electrolyte containing only the inhibitor at various sweeping and RDE rotation rates*

When the rotation rate of the electrode is increased, the global shape of the hysteresis is preserved, but shifted toward more cathodic potentials: it is more difficult to lift inhibition, but also easier to restore it. This is an important observation, as it indicates that the inhibition is lifted in (2) because of some irreversible degradation of the inhibitor. Otherwise, this additive would always be available near the surface, and rotation rate would have little or no effect.

This degradation is compensated by the readsorption of fresh inhibitor from the electrolyte, which is at least partially limited by diffusion, as it is more efficient at high rotation rate. The same explanation holds for the earlier restoration of inhibition during backward sweep.

To summarize, the observed hysteresis during cathodic cycling in the presence of the inhibitor can be interpreted as follows:

1) initially, the inhibitor is strongly adsorbed on the electrode surface;

2) this inhibition holds until the potential is sufficiently cathodic to force Cu^{2+} reduction through the inhibiting adsorbate;

3) once deposition starts, the additive is degraded, probably by incorporation in the film (secondary ion mass spectroscopy (SIMS) analyses, not shown here, reveal a significant organic contamination of Cu);

4) Cu deposition is too fast to allow for the readsorption of the inhibitor;

5) under less cathodic conditions, the Cu deposition slows down and allows for full readsorption of the inhibitor.

Further information can be found in chronopotentiometric measurements (Figure 4.22). As already observed for the suppressor in section 4.2.2.3, upon injection of the inhibitor, the WE is polarized to more cathodic potentials. Upon addition of both the inhibitor and the accelerator, the potential jumps to the same value, indicating that the inhibitor is adsorbed first. However, contrasting with the previous electrolytes, this polarization is not relaxed over time. The deposition remains inhibited, and accelerator does not seem to be able to displace the inhibitor.

4.3.5. The mechanism of extreme fill

Let us now see how this electrolyte performs in TSVs. A typical growth profile obtained with the Gen3 electrolyte in deep TSVs is shown in Figure 4.23. By comparison with Figure 4.17, the difference is striking. With this new electrolyte, an almost perfect bottom-up fill is observed. This regime will be referred to as *extreme fill* [JOS 12].

Based on the results from section 4.3.4, this extreme fill may be interpreted based on the sole action of the inhibitor, considering that:

– its adsorption is limited by mass transport from the electrolyte;

– during electroplating, it is buried in the growing deposit.

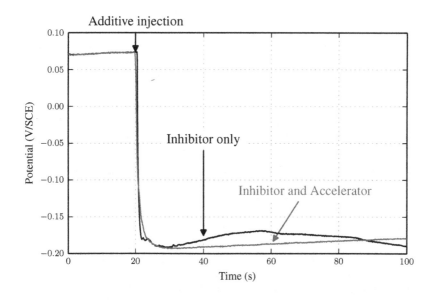

Figure 4.22. *Chronopotentiometric response of inhibitor alone or mixed with accelerator in the Gen3 electrolyte (current density:* $3 \, \mathrm{mA} \cdot \mathrm{cm}^{-2}$*)*

Under these assumptions, the following scenario results (also supported by [JOS 12]):

1) when the substrate is immersed in the electrolyte, the inhibitor almost immediately adsorbs on its surface. However, its penetration into the TSVs is not instantaneous, so a concentration gradient forms within the holes;

2) when electroplating is started, deposition is inhibited where the inhibitor is well adsorbed, but Cu growth may start where this additive is only partially or not adsorbed, i.e. at via bottom;

3) wherever Cu growth has started, reinhibition is not possible, because of the too slow adsorption of the inhibitor: electrodeposition remains active at via bottom and inactive everywhere else.

In the extreme fill regime, the difference in growth rate between bottom and top of the TSVs is huge, almost corresponding to a binary on/off situation[6]. Based on the profiles from Figure 4.23, it can be seen that during the time needed to grow $100\,\mu m$ Cu from the TSV bottom, only 1 to $2\,\mu m$ Cu are deposited outside the hole. Current densities measured during forward and backward cyclic voltammetry (Figure 4.20) at the potential corresponding to the applied current density (about $-200\,mV/SCE$ according to Figure 4.22) are in good agreement with this more than $50/1$ ratio in growth rate.

Figure 4.23. *Deep TSV ($h = 100\,\mu m$, $d = 10\,\mu m$) filled with the Gen3 electrolyte*

---

6 This is not completely true, considering the progressive thickening of Cu observed on the sidewalls in Figure 4.23 during fill. This sloped profile, comparable to the V-shape in Figure 4.17, is a further indication that the adsorption of the inhibitor is limited by mass transport.

Finally, the contribution of the accelerator in this mechanism should be discussed. From the considerations above, this additive is not necessary to obtain extreme fill. However, this small molecule is likely to diffuse faster than the inhibitor, and reach the TSV bottom first. Hence, even if this additive adsorbs slowly, it might be able to compete with the inhibitor. Therefore, it could contribute to further limit inhibition at the via bottom and facilitate the onset of Cu growth at full speed.

4.4. Conclusions

In this chapter, the question of the growth of a new phase, which is probably less discussed than the nucleation in phase transformation, was studied through the example of Cu electroplating for the fabrication of interconnect structures. This process, which aims at filling hollow structures patterned in the substrates, is based on the local modulation of growth rate under the action of additives. More precisely, these additives influence the rate at which cupric ions are able to discharge at the surface of the electrode, which is a key step in the deposition reaction.

The superconformal fill of damascene structures is the spectacular result of the coordinated action of two antagonist additives, the suppressor and the accelerator. If the complex phenomena occurring at the growing surface are probably beyond spatial and temporal resolution of experimental characterization techniques, the essential mechanisms can be approached by the correlation of simple electrochemical measurements with morphological observations. They show that superconformal deposition is the result of (1) the competitive adsorption of the accelerator at the expense of the suppressor and (2) its accumulation within the features due to geometrical contraction during metal deposition. This description provides practical tools to guide the research toward more efficient electrolytes and additive systems, which are constantly needed to properly fill the narrower structures.

More recently, this process was further challenged to fill deep TSVs. In these structures, diffusion-limited replenishment of cupric ions causes additional complication. Nevertheless, void-free filling of these structures has been made possible by the development of a new inhibitor that strongly and irreversibly adsorbs on the surface of Cu to block charge transfer. However, if supersaturation is sufficient, deposition may take over. Therefore, the additive is this time in competition with the deposition reaction itself. This mechanism proves to be more efficient than superconformal growth, as it allows quite larger modulation of the deposition rate between the bottom (full speed deposition) and the bottom (almost no deposition) of the TSVs. This is why it is referred to as extreme fill.

5

Nucleation and Growth of
Metallic Thin Films

In this chapter, we pursue our discussion of nucleation and growth phenomena as key steps in the formation of thin films, i.e. metallic layers that are typically below 100 nm. Indeed, these coatings generally form by nucleation of the isolated islands that grow and coalesce into a continuous deposit, following the principles discussed in Chapter 1. This mechanism is usually referred to as Volmer-Weber [BAU 58]. Early coalescence of the islands is needed to allow the formation of extremely thin continuous films. Hence, a dense enough nucleation is generally required to achieve this goal. This is a central issue in many practical cases, as will be illustrated in this chapter.

This issue will be discussed through the quite detailed description of two processes of interest in the fabrication of the interconnected structures described in Chapter 1: the electrolytic deposition of a thin Cu seed layer and the electroless synthesis of a self-aligned metallic barrier on top of the copper lines.

5.1. Seed layer enhancement

5.1.1. *Electroplating copper seed layers: which electrolyte?*

In the metallization sequence depicted in Figure 1.6, the quality of both the barrier and the seed layer is of paramount importance in order to form void-free and reliable damascene or through silicon via (TSV) structures. In particular, as the AR of the latter increases, the physical vapour deposition

(PVD) deposition of both layers becomes problematic, because PVD is a subconformal deposition technique and allows only partial coverage of vertical surfaces, such as sidewalls [RAD 98]. At some point, it is no longer possible to continuously cover these areas. Consequently, permeable barriers may be formed, and/or voids are left in the metal structures due to incomplete initiation of electroplating (Figure 5.1). In this section, the latter issue is addressed and a complementary process to PVD is introduced, aiming at electroplating Cu onto surfaces where the barrier material is exposed (i.e. places where seed layer layer is absent). This process will be referred to as seed layer enhancement (SLE) [CUZ 10].

Figure 5.1. *Shallow TSV ($h = 17 \, \mu m$, $d = 3 \, \mu m$) seeded with $50 \, nm$ PVD Cu tentatively filled by Cu electroplating. In the lower part of the via, the PVD seed layer layer was not continuous, preventing initiation of Cu electroplating*

The requirements for the SLE process differ from those of the Cu electroplating processes discussed in Chapter 4. Indeed, this process aims at:

1) *conformally* depositing thin layers (typically 20 to 100 nm);

2) onto resistive substrates (such as exposed – and possibly oxidized – barrier material).

Both requirements impose a drastic decrease in bath conductivity by comparison with Cu fill electrolytes without seriously reducing Cu content. This imposes a decrease in acid content (which is the supporting electrolyte in Cu fill electroplating solutions), increasing the pH [KIM 07]. Unfortunately, cupric ions are not stable in solution when the pH exceeds 5–7 (Figure 5.2) [POU 63]. Indeed, above these values, solid oxides and hydroxides form.

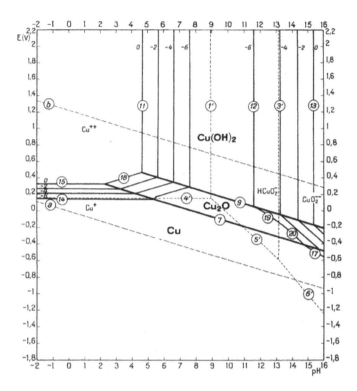

Figure 5.2. *Pourbaix diagram of Cu [POU 63]*

To stabilize a soluble form of Cu(II) in the neutral or alkaline pH range, it is necessary to add an appropriate complexing agent. For SLE, ethylenediamine (EN) is added into the solution with a 2 : 1 ratio with respect to Cu^{2+}. The corresponding formulation is detailed in Table 5.1. Not mentioned in this table is boric acid, which is present in the bath as a buffering agent. The pH is fixed by the addition of appropriate amounts of tetramethylammonium hydroxide (TMAH). The corresponding Pourbaix diagram has been computed for this electrolyte using the JChess software (Figure 5.3). In this diagram, a domain is clearly visible in the pH 4–12 range in which metallic Cu is in equilibrium with dissolved $Cu(EN)_2^{2+}$ complexes. In this domain, the electrode position of Cu from $Cu(EN)_2^{2+}$ is thus possible through the reaction:

$$Cu(EN)_2^{2+} + 2\,e^- + 2\,H^+ \rightleftharpoons Cu + 2\,HEN^+ \qquad [5.1]$$

The Nernst potential can be directly read on the diagram at the corresponding pH, and is given in Table 5.1.

	Composition		pH	Nernst potential
	[Cu2+] (mol/L)	[EN] (mol/L)		(V/SCE)
SLE	4×10^{-2}	8×10^{-2}	9.5	-0.47

Table 5.1. *Formulation of the SLE electrolyte*

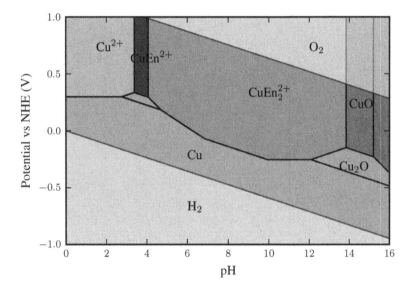

Figure 5.3. *Pourbaix diagram of Cu in the SLE electrolyte. For a color version of this figure, see www.iste.co.uk/haumesser/metals.zip*

5.1.2. First implementation of the SLE process in shallow TSVs

In a first approach, let us examine the restoration of the discontinuous 50 nm-thick PVD seed layer in shallow TSVs (Figure 5.1) by electroplating Cu from a SLE electrolyte. In Figure 5.4, the deposit obtained using a pulsed current waveform (10 ms pulses at $-10\,\mathrm{mAcm^{-2}}$ followed by 90 ms without current) is compared to the corresponding poor fill result displayed in Figure 5.1. From these figures, it is clear that:

– where the PVD seed layer is continuous (in the upper part of the TSV), the SLE deposit is also continuous and smooth;

– where the PVD seed layer is discontinuous (in the lower part of the TSV), the SLE deposit is also discontinuous, and exhibits a nodular morphology.

The latter morphology is characteristic of too sparse a nucleation. As discussed in Chapter 2, this electroplating reaction proceeds by heterogeneous nucleation of Cu onto the barrier material. The Cu islands then grow and eventually coalesce into a continuous film. If nucleation is too sparse, the distance between neighboring Cu islands is too large to allow proper film closure, leading to the typical morphology in Figure 5.4 (left).

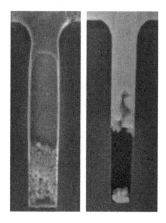

Figure 5.4. Left: Shallow TSV (h =17 μm, d =3 μm) seeded with 50 nm PVD Cu tentatively repaired by SLE. In the lower part of the TSV, where the PVD seed layer was not continuous, a nodular deposit is observed. Right: For comparison, the same TSV after Cu fill, without SLE (same as Figure 5.1), shows the good correspondence of the bottom void with the area of nodular SLE

5.1.3. Nucleation of SLE on resistive barrier materials

This nucleation step can be studied on flat electrodes using conventional electrochemical techniques (see Chapter 3). Figure 5.5 shows the cyclic voltammograms obtained for polished Si samples covered with 20 nm TiN or a more conventional TaN/Ta bilayer barrier (total thickness 25 nm).

Cu deposition does not start immediately below the Nernst potential of SLE ($-0.47\,\text{V/SCE}$, see Table 5.1) on either material. *An overpotential needs to be applied to initiate the reaction and measure a current.* This is typical of a situation in which Cu has to nucleate first and is expected on heterogeneous materials such as the barrier layers (see section 3.1.3).

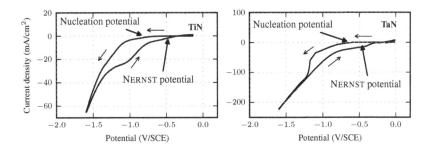

Figure 5.5. *Cyclic voltammetry in the cathodic region for the SLE process on TiN (left) and TaN/ Ta (right), showing that deposition requires a nucleation overpotential with respect to the Nernst potential*

To get further insight into this matter, the scanning electron microscope (SEM) image of a TiN sample treated by SLE using the pulsed current waveform described in section 5.1.2 is shown in Figure 5.6. From this image, nucleation density can be estimated to $9 \times 10^9\,\text{cm}^{-2}$ only . This corresponds to an average distance of $100\,\text{nm}$ between nuclei, which is clearly excessive considering the targeted thickness for the SLE layer (typically less than $200\,\text{nm}$).

Overall, these results indicate that nucleation is fairly difficult on both barrier materials, as these surfaces lack active nucleation sites.

5.1.4. Enhancement of nucleation density on resistive barriers

Both TiN and Ta are known to spontaneously oxidize in air [RAD 04]. The presence of a superficial oxide could limit Cu nucleation by:

Figure 5.6. *SEM picture of an SLE deposit on* $20\,\mathrm{nm}$*-thick TiN – total charge density* $20\,\mathrm{mC} \cdot \mathrm{cm}^{-2}$ *(corresponding to* $7\,\mathrm{nm}$ *nominal thickness) using the pulsed current waveform depicted in section 5.1.2*

– screening the charge transfer needed to reduce cupric ions into metallic copper[1];

– modifying the energetic cost of formation of the Cu/substrate interface and increasing the work of nucleation;

– limiting the mobility of adatoms (see section 3.1.1.3);

– finally, by blocking nucleation sites on the barrier surface.

It is thus interesting to verify whether upon removal of this oxide, the nucleation density of Cu is improved. Indeed, this oxide can be (at least partially) converted back to a metallic state by cathodic electrolysis in a supporting electrolyte. The latter is a solution of boric acid at a pH of 9.5. The treatment is carried out well beyond the potential of water reduction to allow for the reduction of Ti or Ta oxides back to metal.

The impact of this treatment on the surface of the barrier materials can easily be assessed using water contact angle measurements (Figure 5.7). Upon treatment, the contact angle drastically decreases, indicating that the barrier materials become more hydrophilic, as expected for metallic surfaces.

---

1 Indeed, when comparing Figure 5.5 with Figures 4.5 and 4.19, it is interesting to note that the addition of a suppressor or an inhibitor in a Cu electrolyte also results in an overpotential for Cu deposition, which could be ascribed to difficult nucleation of the deposit through the inhibiting layer.

Over time, however, the contact angle increases again toward its initial value, indicating that the surfaces are progressively reoxidized in air.

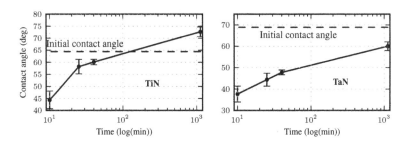

Figure 5.7. *Evolution of water contact angle over time after electrochemical deoxidation of TiN (left) and TaN/Ta (right) barriers, compared with its initial value before treatment. The increase in contact angle over time is attributed to progressive reoxidation of the surfaces under air*

Oxygen profiles measured by secondary ion mass spectrometry (SIMS) analysis for both barrier materials coated with SLE Cu are shown in Figure 5.8. For each barrier material, oxygen profiles with and without surface deoxidation are compared. For both pristine TiN and TaN/Ta, a peak in oxygen content is measured right below the interface of the barrier material. Upon deoxidation, this peak is significantly attenuated, showing lower oxygen content in the stack. This gives clear evidence that the deoxidation process actually (at least partially) removes the surface oxide, which is converted back to metal (Ti or Ta) by this cathodic treatment.

The impact of this treatment on SLE deposition kinetics can be verified by cyclic voltammetry, following the procedure described in section 5.1.3. In Figure 5.9, the voltammograms of deoxidized samples are compared to the initial ones (same as in Figure 5.5). For both barriers treated, the forward sweep now coincides with the backward sweep, showing good reversibility of the system. The nucleation overpotential has almost completely vanished. This confirms that the superficial oxide was probably blocking the nucleation process, and that the deoxidation treatment efficiently restores the barrier surface and facilitates nucleation. Finally, the improved nucleation of SLE Cu is confirmed by SEM (Figure 5.10). This shows that the deposit is composed of metallic (presumably Cu) clusters with two populations:

– large clusters (30–50 nm) whose density is about $1 \times 10^{10}\,\text{cm}^{-2}$;

– small clusters (a few nm in diameter) that almost entirely cover the surface (density estimated to be $4.5 \times 10^{11} \, \text{cm}^{-2}$).

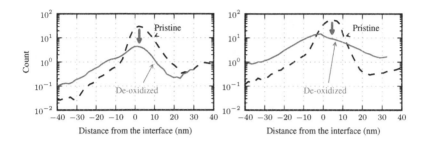

Figure 5.8. *Oxygen profiles measured by SIMS near the TiN/Cu interface (left) and Ta/Cu interface (right). Dashed lines: SLE Cu deposited onto pristine barriers. Solid lines: SLE Cu deposited onto deoxidized barriers. Intensities have been normalized with respect to Cu. In both cases, oxygen content at the interface is decreased for treated barriers*

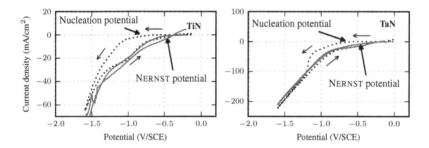

Figure 5.9. *Cyclic voltammetry in the cathodic region for the SLE process on TiN (left) and TaN/Ta (right), without (dotted line, corresponding to Figure 5.5) and with (solid line) deoxidation treatment. When barriers are deoxidized, the nucleation overpotential vanishes*

The reason two populations are formed is not clear. However, the overall density of Cu clusters is much higher than in Figure 5.6, which confirms the more efficient Cu nucleation.

Figure 5.10. *SEM picture of SLE deposit on* deoxidized $20\,\mathrm{nm}$ *thick TiN – total charge density* $20\,\mathrm{mC} \cdot \mathrm{cm}^{-2}$ *(corresponding to* $7\,\mathrm{nm}$ *nominal thickness) using the pulsed current waveform described in section 5.1.2*

All experimental results thus converge to demonstrate that an electrochemical cathodic treatment of the TiN and TaN/Ta barrier surfaces is efficient at removing the superficial oxide formed upon air exposure. This restores the metallic character of the surfaces, which become more active toward Cu nucleation during the SLE process.

5.1.5. *Integration of the SLE process*

A shallow TSV repaired by the SLE process implementing the deoxidation step is shown in Figure 5.11(b). The field of the substrate is covered with a $80\,\mathrm{nm}$-thick copper film. Inside the TSV, this thickness decreases to $45\,\mathrm{nm}$. Below the limit where the PVD seed layer is not continuous, the deposit unfortunately remains nodular, as shown in Figure 5.4. However, the Cu nodules are smaller, which indicates a more active nucleation, in line with the results of the study in section 5.1.4.

In order to further increase nucleation at the bottom of the via, another possibility is to increase supersaturation, i.e. to impose a more cathodic potential. As the SLE process is current controlled, this is done by increasing the cathodic peak current density during pulsed deposition. When this current density is increased from $10\,\mathrm{mA} \cdot \mathrm{cm}^{-2}$ to $20\,\mathrm{mA} \cdot \mathrm{cm}^{-2}$ (Figure 5.11(c)), a similar morphology as shown in Figure 5.11(b) is observed, with a nodular deposit composed of small Cu particles at the bottom of the via. Nevertheless, the limit between this zone and the continuous deposit has shifted deeper into the TSV. The Cu layer thus remains continuous down to $15\,\mu m$ within the via, where it is about $40\,nm$ thick. The remaining $2\,\mu m$ at the bottom of the via may suffer from an excessive ohmic drop and/or Cu(II) ion depletion caused by the higher current density. Indeed, both phenomena could contribute to a decrease in the supersaturation in this area.

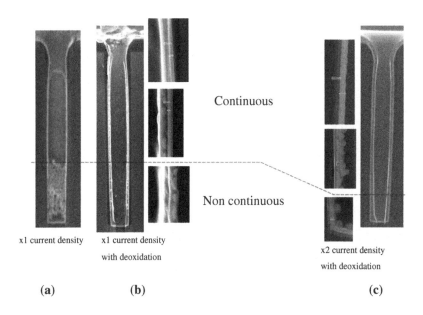

Continuous

Non continuous

x1 current density   x1 current density

with deoxidation

x2 current density

with deoxidation

(a)   (b)   (c)

Figure 5.11. a) Shallow TSV (h $=17\,\mu m$, d $=3\,\mu m$) seeded with $50\,nm$ PVD Cu tentatively repaired by SLE (same as Figure 5.4 (left)); b) the same TSV repaired by SLE , but including intermediate deoxidation; c) the same TSV repaired by deoxidation and SLE at a higher cathodic peak current density $(20\,mA \cdot cm^{-2})$

In spite of the still unfavorable morphology of the SLE deposit, this process can be integrated within the metallization sequence, including Cu fill (using the first-generation electrolyte described in section 4.3.3, see Figure 4.17) and CMP. As can be seen in Figure 5.12, even if the SLE is not perfectly smooth at the bottom of the via, it nonetheless enables a void-free electroplating fill.

The electrical resistance of TSVs metallized with the SLE process can be measured in specific structures, called Kelvin vias. These structures are designed to measure the resistance of a single via using a 4-point probe technique. To assess the performance of the SLE process, two different metallization sequences are compared:

1) A test sequence:

   i) 20 nm TiN deposited by PVD,

   ii) 50 nm Cu deposited by PVD using the same equipment, without an air break [MAI 04],

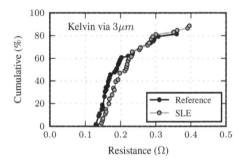

Figure 5.12. *Left: Morphology of a shallow TSV (h =17 µm) after full metallization sequence, including deoxidation and 100 nm SLE Cu. Right: Corresponding electrical performances of Kelvin structures, compared to the reference metallization (in which the thin PVD liner is covered with a 150 nm thick Cu chemical vapor deposition (CVD) film)*

   iii) *cathodic deoxidation,*

   - iv) 100 nm *Cu deposited by SLE,*

   v) Cu fill,

   vi) CMP.

2) A reference sequence:

i) 20 nm TiN deposited by PVD,

ii) 50 nm Cu deposited by PVD using the same equipment, without an air break,

iii) 150 nm *Cu deposited by CVD using the same equipment, without an air break,*

iv) Cu fill,

v) CMP.

The corresponding results are plotted in Figure 5.12 (right). This is a cumulative plot, where 100 % corresponds to 55 Kelvin vias. The fact that any of the two curves reaches 100 % reveals that some structures are electrically open (i.e. they exceed the acceptable resistance threshold, which is fixed at 300 mΩ). The remaining functional structures correspond to the *yield* of the process. For both metallization sequences, a similar yield of 80–90 % is obtained. The other relevant information is the median resistance value of the two populations. For the reference process, this median resistance is 180 mΩ. When the TSVs are fabricated using the SLE process, a similar value of 210 mΩ is obtained.

Finally, the SLE process with enhanced nucleation (by surface deoxidation and increased supersaturation) performs similarly to a reference seeding sequence in terms of final morphology and the electrical characteristics of the TSVs.

5.2. Electroless deposition of self-aligned metallic barriers

5.2.1. Why would aligned metallic barriers be needed?

As depicted in Figure 1.6, the ultimate step in the fabrication of a damascene metal level is the encapsulation of the Cu lines. This step is of paramount importance, as this encapsulation layer participates in the confinement of Cu within the interconnection lines; the metal would otherwise be in direct contact with the dielectric insulator from the next metal level. However, this layer is also a source of defects in the interconnect structures.

Usually, these encapsulation layers are formed with dielectric materials such as SiN, SiCN or derivatives [LE 13]. By comparison with the interline dielectric materials introduced in section 1.1.2, these films exhibit a rather larger dielectric constant, and are major contributors in the capacitance of the interconnects [CHA 07]. Therefore, their elimination would reduce the resistive capacity delay in these structures (see section 1.1.1).

More importantly, the interface between Cu and these dielectric barriers has been shown to be the weak point regarding EM [LAN 03]. This phenomenon is responsible for the aging and eventually the failure of the interconnect structures. EM is the drift of Cu atoms caused by the flow of current (current densities of about $1\,\mathrm{MA} \cdot \mathrm{cm}^{-2}$ flow through the narrowest metal lines) combined with thermal activation (typically, the temperature of the interconnect structures exceeds $100\,°\mathrm{C}$ under the normal operation of a computer chip). Recent studies have shown that these movements of matter nucleate voids[2] at the Cu/dielectric barrier interface [LLO 91], which aggregate near the vias (where mechanical stress is concentrated) and finally disrupt the interconnections [LIN 02]. As these structures are miniaturized, the local peak current density under normal operation increases. EM is enhanced, causing early failure of the whole device.

As the circuit disruption by EM is activated by current and temperature, *accelerated aging tests* are usually performed at high current and temperature. In these tests, EM failure has been shown to obey an empirical, Arrhenius-like law, called *Black's equation* [DWY 10], which expresses the mean time to failure (MTTF) of a population of chips as:

$$\mathrm{MTTF} = A J^{-n} \exp\left(\frac{E_a}{kT}\right) \qquad [5.2]$$

where A, n and E_a are parameters of the model. Among these parameters, the activation energy E_a plays a central role (as it comes in the exponential term) in extrapolating the overall resistance of the structure toward EM *under normal operating conditions*. The higher this energy, the more durable the conducting lines and vias. For typical stacks involving dielectric barrier materials, this activation energy is about 0.8–$0.9\,\mathrm{eV}$, and corresponds to the nucleation of voids at the Cu/dielectric barrier interface [LIN 02]. Unfortunately, for more advanced devices, this activation energy is not

---

2 This is another example of nucleation.

sufficient to allow for their durability to reach the target value of 10 years under normal operating conditions.

To improve durability, we thus need to remove the weak Cu/dielectric barrier interfaces. Therefore, it has been proposed we replace this interface with a Cu/metal one (Figure 5.13) [GAM 06]. However, this supposes we are able to *align* this metallic barrier above copper lines; otherwise, this layer would create short circuits between adjacent conductors. This alignment could be done by additional patterning steps including well-aligned lithography. However, this would be impractical, expensive and technically complex. A more elegant approach consists of *selectively depositing* this barrier on top of the copper lines to form the so-called self-aligned barriers (SABs) [SHA 03]. The process capable of such a selective reaction is *electroless deposition* and is described hereafter.

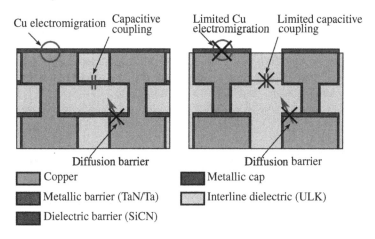

Figure 5.13. *Left: Conventional capping of Cu lines by uniform dielectric barriers. They increase capacitive coupling and form a weak interface with Cu, which is a preferential location for the formation of EM-related defects. Right: The alternative strategy using aligned metallic barriers aims at addressing both issues. For a color version of this figure, see www.iste.co.uk/haumesser/metals.zip*

5.2.2. *Principles of electroless deposition*

In electroless processes, the reaction of metal deposition is the same as in electroplating, i.e. the reduction of metal ions into zero valent metal according to equation [3.1]. Therefore, all basic principles of metal electrodeposition hold in electroless processes. The fundamental and only difference with electroplating is that the electrons in equation [3.1] are not supplied by an external source, but by an oxidation reaction concomitant with the reduction of the metal ions. As a result, the electroless reaction is a balanced *redox* process composed of two half-reactions:

$$Me^{z+} + ze^- \longrightarrow Me$$

$$Red \longrightarrow Ox + ne^-$$

$$nMe^{z+} + zRed \longrightarrow nMe + zOx \qquad [5.3]$$

5.2.2.1. *Thermodynamic considerations*

For this reaction to actually occur, the initial state (nMe^{z+} + zRed) must be less stable than the final state (nMe + zOx). In such redox systems, this means that the equilibrium potential of the Me^{z+}/Me couple must be higher than that of the Ox/Red couple (see Figure 5.14). A non-exhaustive list of suitable reducing agents and metals is given in Table 5.2.

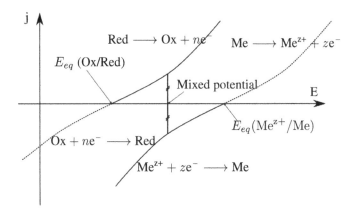

Figure 5.14. *Schematic representation of the mixed potential theory*

If it were only for thermodynamics, a solution containing metal ions and a reducing agent should spontaneously precipitate the metal. This reaction can be understood by the *mixed potential theory* illustrated in Figure 5.14. Both half-reactions in equation [5.3] possess their own current/potential characteristic (as measured by cyclic voltammetry, for instance). Due to the arrangement of equilibrium potentials, there is a domain of potential where both reactions coexist. In this domain, there is a specific potential at which both reactions occur at the same rate (i.e. at the same current density). This potential is called *mixed potential*, and the common current values correspond to the metal precipitation rate (as deduced from Faraday's law, equation [3.23]).

	Reactant	Redox couple	E^0 (V/standard hydrogen electrode)
Metal	Gold	Au^{3+}/Au	1.50
	Silver	Ag^+/Ag	0.80
	Copper	Cu^{2+}/Cu	0.35
	Nickel	Ni^{2+}/Ni	-0.25
	Cobalt	Co^{2+}/Co	-0.28
Reducing agent	dimethyleamineborane (DMAB)	$BH_3/H_2BO_3^-$	-1.2
	Sodium hypophosphite	$H_2PO_2^-/HPO_3^{2-}$	-1.4

Table 5.2. *Metals and reducing agents in electroless processes*

5.2.2.2. *Kinetic control of electroless reactions: catalytic activation*

This precipitation could proceed by homogeneous or heterogeneous (on the reactor's walls) nucleation and growth of the metal, but this is not our aim here. Instead, kinetic control of the reaction is required to:

1) avoid undesired metal deposition (between Cu lines on the substrate, but also on the reactor's wall, in the tubing, etc.);

2) activate deposition wherever it is needed (on top of the Cu lines).

Such a *selective* deposition can be achieved by appropriate formulation of the solution. The following aspects can be considered to achieve this:

– the relative and absolute concentrations of metal salt and reducing agent;

– the complexation of the metal ions[3];

– the temperature;

---

3 In the Nernst relation equation [3.11], it is the actual concentration of Me^{z+} which must be used. Upon complexation, this concentration can become extremely low, depending on the equilibrium constant of the complexation reaction. Therefore, the Nernst potential is always shifted to more negative values upon complexation of metal ions.

– the pH;

– the addition of stabilizing agents.

As defined in Chapter 2, a state in which a system is blocked above its minimal Gibbs energy is called *metastable*. This is the case for electroless solutions.

The beauty of electroless processes resides in the ability to *activate* the redox reaction on *catalytic* surfaces. Indeed, certain metals are known to be able to accelerate chemical reactions without being consumed in the process, which is the definition of a *catalyst* [FRE 11]. In the case which is of interest to us, the catalyzed reaction is the oxidation of the reducing agent [OHN 85]. This reaction is

$$H_2PO_2^- + H_2O \longrightarrow HPO_3^{2-} + 3\,H^+ + 2\,e^- \qquad [5.4]$$

for hypophosphite and

$$BH_3 + 3\,H_2O \longrightarrow H_2BO_3^- + 7\,H^+ + 6\,e^- \qquad [5.5]$$

for dimethyleamineborane (DMAB). Generally, in electroless processes, the deposited metal is catalytic towards these reactions [OHN 85]. For this reason, the electroless process is called *autocatalytic*. This means that once started, the reaction does not stop, and the metal deposit continues to grow.

To localize the electroless deposit, we therefore need to make sure that the desired location is catalytic. Interestingly enough, Cu catalyzes the oxidation of borane derivatives such as DMAB [OHN 85]. Upon dipping entire wafers in such solutions, deposition is thus expected to start only on top of the copper lines. From a practical standpoint however, these reducing agents are usually less stable than hypophosphite. The latter is thus generally preferred to formulate longer lasting electroless baths. Unfortunately, Cu is not a good catalyst of the oxidation of hypophosphite. This reaction is catalyzed by Pd. Luckily, this metal is more noble than Cu. Therefore, upon immersion of metallic copper in a Pd^{2+} solution, a *displacement reaction* takes place:

$$Pd^{2+} + Cu \longrightarrow Pd\downarrow + Cu^{2+} \qquad [5.6]$$

Therefore, a Pd deposit spontaneously forms at the surface of the Cu lines, and – hopefully – *nowhere else*. This process is called *activation* of the Cu surface. A tiny Pd deposit – a fraction of a monolayer – is sufficient to catalyze the electroless reaction and create the nucleation sites for barrier deposition.

By taking advantage of the catalytic properties of Cu (for borane derivatives), or by selectively activating its surface with a tiny Pd deposit (for hypophosphite), the electroless reaction can be kinetically restricted to the surface of metal lines, yielding selective deposition.

Finally, it should be mentioned that in addition to the electroless reaction in equation [5.3], a parasitic reaction always occurs: the *reduction of the reducing agent*. Indeed, in the case of hypophosphite, its oxidation [5.4] forms H^+ ions, which locally acidify the solution and allow the following reaction [POU 63]:

$$H_2PO_2^- + 2\,H^+ + e^- \longrightarrow P + 2\,H_2O \qquad [5.7]$$

Similarly, the oxidation of borane derivatives also decrease the pH locally, which allows the reduction of the oxidized product into boron:

$$H_3BO_3^- + 4\,H^+ + 3\,e^- \longrightarrow B + 3\,H_2O \qquad [5.8]$$

Because of the parasitic reaction of the reducing agent, elemental B (for borane derivatives) or P (for hypophosphite) is always codeposited with the metal during electroless reaction.

5.2.3. *Experimental conditions for electroless deposition*

5.2.3.1. *Selection of a metal*

Among the metals listed in Table 5.2, silver and gold cannot be used to form barriers against Cu diffusion, as they are miscible with this metal. Upon thermal stress during the fabrication or operation of the devices, they are expected to diffuse into the Cu lines and cause a dramatic increase in electrical resistance. In addition, gold is a known contaminant of silicon devices.

Nickel also is miscible with Cu. Nonetheless, it has been considered a candidate for the fabrication of SABs [CHO 08, ANT 06]. Oddly, the expected resistance increase of the line has not to our knowledge been reported with these barriers.

Finally, the most frequently studied metal is without doubt cobalt [ALM 07, GAM 06]. This metal is not miscible with Cu, and should thus form an efficient barrier. But even in this case, Cu diffusion at the grain boundaries of a polycrystalline Co(P) or Co(B) film cannot be discarded. To limit this phenomenon, a third metal is usually added into the film. Refractory metals are reputed to efficiently stabilize the amorphous microstructure of the layers and have a very high thermal stability [ANT 06]. Mo, Rh and W have been studied [SHA 03, GAM 06]. The latter has attracted much interest. By the addition of a tungstate salt in the electroless solution, W is co-deposited [EIN 05]. It should be noted that tungstate ions are only stable in alkaline media [POU 63]: for this reason, the electroless solutions used for the fabrication of SAB have a pH above 9.

Here, two electroless processes will be considered:

– a CoWP process using hypophosphite as a reducing agent, and requiring adequate activation of the Cu surfaces;

– a CoWB process, using a borane derivative as a reducing agent, which is capable of direct deposition onto the Cu lines.

5.2.3.2. *Composition of electroless solutions*

The various components of a typical electroless solution for CoWB deposition are listed in Table 5.3. Typical operation conditions for both CoWP and CoWB electroless solutions are given in Table 5.4.

Role	Component	Compound	
Metal source	Metal salt	Cobalt sulfate	$CoSO_4 \cdot 7\,H_{20}$
	Complexing agents	Sodium citrate	$C_6H_5Na_3O_7 \cdot 2\,H_2O$
		Malic acid	$C_4H_6O_5$
Electron donor	Reducing agent	Morpholine borane	$C_4H_9ON-BH_3$
Refractory metal	Salt of refractory metal	Sodium tungstate	$Na_2WO_4 \cdot 2\,H_2O$
Stabilizer	Additives	Surfactants	
pH control	Acids/bases		TMAH

Table 5.3. *Composition of the CoWB solution*

Process	pH	Temperature (°C)
CoWP	9 to 10	70 to 80
CoWB	9 to 10	50 to 60

Table 5.4. *Operation parameters of the CoWP and CoWB processes*

5.2.3.3. Process flow for the deposition of CoWP and CoWB

As a whole, the process flow for electroless deposition is quite complex. All the steps listed in Figure 5.15 have been introduced before, except the first one, which is called *pre-clean*. This step is of crucial importance for the successful deposition of the SAB. Indeed, this process directly follows CMP, which leaves a contaminated Cu surface. This contamination is mostly deliberate: in the last step of the CMP process, the substrates are exposed to organic compounds such as benzotriazole, which saturate the surface of the Cu lines to prevent their corrosion during wafer storage. Obviously, these compounds are also efficient at inhibiting the surface reactions involved in the electroless process. Therefore, prior to SAB deposition, these contaminants and other CMP residues need to be eliminated by adequate cleaning solutions.

In industrial equipment, the surface preparation steps (pre-clean and activation) are generally performed in spray chambers. For the deposition reaction, immersion is preferred, as it allows for improved control of temperature and local pH (see Table 5.4).

5.2.4. Deposition kinetics and film composition

The deposition rate can be measured for both processes using bare Si wafers uniformly coated with Cu and polished following the same CMP process as structured substrates. The Co alloy thickness is typically measured by X-ray fluorescence (XRF), previously calibrated by absolute X-ray reflectometry measurements. XRF is a faster technique that allows the thickness to be mapped across entire wafers. The evolution of thickness over time is plotted in Figure 5.16 for CoWP and CoWB.

In both cases, linear growth of the films is observed. Interestingly enough, the curves do not intercept the origin. Instead, some incubation time is needed before the films can be detected. This corresponds to the time lag for nucleation of the CoWP and CoWB alloys, as described in section 2.2.6.1.

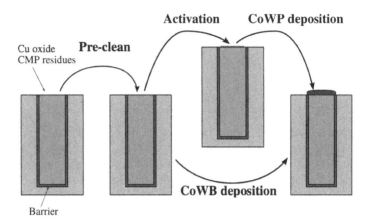

Figure 5.15. *Schematic representation of the process flow for electroless deposition*

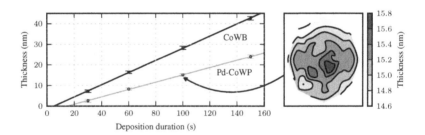

Figure 5.16. *Evolution of film thickness with deposition duration for the CoWP and CoWB processes, and example of a thickness map of a* 15 nm *thick CoWP deposit. Error bars correspond to* ±2 *standard deviations of thickness across the wafer*

It is also remarkable that under the accurate pH and temperature control allowed by an industrial reactor, the variation in thickness across the wafer remains below 2 %, which for a 15 nm-thick deposit corresponds to a peak-to-valley variation of less than 1 nm.

The composition of the CoWP and CoWB films is reported in Table 5.5. A first noticeable result is that the hypophosphite-based process leads to a fair amount of P in the film, whereas B is hardly present in CoWB deposits. Conversely, W content is much higher in CoWB.

Material	Co	W	P	B	Crystal structure
CoWP	91	1	8		Well crystallized (hexagonal close-packed (hcp))
CoWB	90	9		1	Crystallized (hcp and face-centered cubic (fcc))

Table 5.5. *Composition (in %) and crystalline structure (determined by X-ray diffraction) of the CoWP and CoWB films*

5.2.5. Local thickness variations

Although thickness control is remarkable at the wafer scale, atomic force microscopy (AFM) and SEM observations in Figure 5.17 reveal significant variations at the microscale. Both materials exhibit a granular morphology with typical grain size of a few tens of nanometers, which is the result of the nucleation, growth and coalescence of discrete islands. Clearly, this morphology is variable between large domains, with a typical extension of several hundred nanometers. In the AFM profile of a CoWP film in Figure 5.17 (left), domains with straight boundaries are revealed, in which the cobalt deposit appears to be significantly thinner and rougher. This behavior is also observed on interconnect structures (Figure 5.17 (right)). A similar morphology is observed for self-initiated CoWB films (not shown here). Further examination suggests that these domains correspond to places were nucleation is less dense. Indeed, as suggested in section 5.1.4 for the deposition of a Cu seed layer, the nucleation density conditions the thickness needed to coalesce and close the film (the denser the nucleation, the earlier the coalescence) and in turn affects the final roughness (less dense nucleation leads to larger islands before coalescence, thus rougher deposits).

These morphological variations in the electroless films could compromise their functional properties and compatibility with the subsequent integration steps. For instance, excessive roughness associated with sparse nucleation might affect the fabrication of the upper metal level. More importantly, for thin coatings compatible with the 32 nm node or below (i.e. with a nominal thickness of 8 nm or less), it is expected that the domains with less dense nucleation will not be properly covered. If the cobalt-based films are not continuous, surface diffusion of copper will not be completely blocked, leaving fast pathways for EM. For this reason, it is necessary to determine the origin of these variations. This is the purpose of the following section.

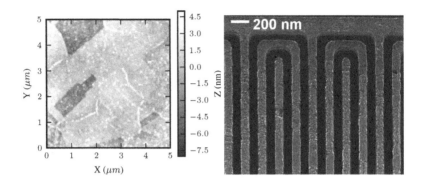

Figure 5.17. *Left:* *AFM image of a* 30 nm-*thick CoWP film deposited on polished copper.* *Right:* *SEM observation of a small area of a* 90/90 nm-*width/space comb/serpentine structure coated with a* 10 nm-*CoWP film. For a color version of this figure, see www.iste.co.uk/haumesser/metals.zip*

5.2.6. *Factors influencing the nucleation of CoWP and CoWB*

5.2.6.1. *Nucleation of palladium on copper*

In a first step, let us examine the nucleation of the Pd activation layer on copper. This reaction creates the nucleation sites for the CoWP layer. Therefore, any difference in the density of Pd nucleation across the copper surface is expected to translate into the subsequent CoWP deposit. In the SEM image in Figure 5.18, the palladium clusters appear as clear dots against the darker background of the copper surface. This picture reveals the spectacular arrangement of these clusters: domains with straight boundaries are visible, in which denser Pd nucleation is obvious. This observation corroborates the revelation of similar domains after electroless deposition of CoWP films (Figure 5.17 (left)).

These straight boundaries are likely to correspond to twin boundaries within the copper grains. In turn, this would mean that palladium nucleation is highly sensitive to the crystallographic orientation of the underlying copper surface. This is proved by electron backscatter diffraction (EBSD) mapping of the same area (Figure 5.18 (right)).

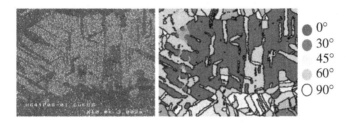

● 0°
● 30°
45°
● 60°
○ 90°

Figure 5.18. *Left: SEM image of the Pd nuclei on the polycrystalline copper surface. Right: Corresponding EBSD orientation mapping of the copper grains (angles refer to disorientation with the (111) direction), showing strong correlation between dense Pd nucleation and Cu(111) orientation. Adapted from [OLI 10]. For a color version of this figure, see www.iste.co.uk/haumesser/metals.zip*

The correlation between the areas of dense palladium nucleation and the copper grains with (111) orientation normal to the surface is clearly visible, confirming that copper orientation plays a significant role in the electroless nucleation mechanism.

Considering only the electrochemical and surface properties of the Cu(111) surface, this result could be surprising. These dense planes of the cubic structure of copper are the most noble electrochemically speaking (with a work function of $4.9\,\mathrm{eV}$ [REN 06]), and should exhibit a lower density of nucleation sites (defects) than other crystallographic planes. For both reasons, we would expect less palladium nucleation on these planes as compared to other directions. However, these more stable planes are also less likely to dissolve during the activation reaction. As a result, they concentrate cathodic sites (where Pd is deposited) at the expense of anodic sites (where Cu is dissolved), which may be located on less stables areas. As all Cu grains are electrically connected, cathodic and anodic sites are not necessarily close to each other.

Another explanation involves the crystallographic properties of both metals. Palladium crystallizes according to the same face-centered cubic (fcc) structure as copper. Furthermore, their lattice parameters are close, with a misfit of less than $8\,\%$. Hence, it has been reported that in ultra-high vacuum processes such as molecular beam epitaxy, palladium grows pseudomorphically on Cu(111) [PAN 04]. Symmetrically, a similar mechanism is reported for the deposition of copper onto Pd(111) [DE 05]. If this interfacial behavior holds in our chemical deposition, it is expected to minimize the energetic cost of the interface and favor palladium deposition onto the Cu(111) surface. This could contribute to the spectacular decrease in nucleation density between the (111)-oriented grains and their highly misoriented lateral twins. Among the four twin variants of a (111)-oriented grain, only one variant preserves the (111) orientation (but this variant is below the observed the surface) and the three lateral variants with the visible straight twin boundaries are (221)-oriented (see, for example, [CAY 07] for more details on twins).

5.2.6.2. *Crystallographic structure of the Co/Cu interface*

The results obtained for Pd deposition cannot explain the Cu grain dependence of self-initiated CoWB coatings, as Pd activation is not involved in this process. In this case, differences in interfacial energies are the more likely cause of variations in nucleation density. A transmission electron microscopy (TEM) cross-section and microdiffraction patterns of a Cu/CoWB interface are displayed in Figure 5.19. The irregular surface of the deposit is clearly evident, in agreement with the SEM observation in Figure 5.17. The polycrystalline microstructure of the underlying copper is also visible, and an individual Cu grain can easily be identified as a darker area close to the interface with the CoWB layer. The microdiffraction pattern of this area (Figure 5.19(c)) reveals that this domain is crystallized in the classical copper fcc system. The second microdiffraction pattern from the CoWB film above this particular Cu grain (Figure 5.19(b)) reveals that the CoWB material is also crystalline, with a fcc structure. This crystallographic structure does not correspond to the stable hexagonal close-packed (hcp) structure of pure Co. In addition to the identification of the lattice systems of both copper and CoWB, the microdiffraction patterns in Figure 5.19 show perfect alignment of the orientation of both domains. *There is thus a complete epitaxy between the CoWB film and copper grains.* This result is supported by the small lattice misfit between the two structures, which is less than 2 %.

A similar observation for CoWP is displayed in Figure 5.20. Again, the fcc structure of copper is clearly identified from the microdiffraction pattern in Figure 5.20(c). The CoWP film exhibits a well crystallized columnar microstructure. The electron diffraction pattern in Figure 5.20(b) reveals that in this case the cobalt alloy crystallizes along the hcp structure. However, there is still a clear orientation relationship between the copper and hcp cobalt grains: $(111)_{Cu} \parallel (0001)_{Co}$ and $(\bar{4}22)_{Cu} \parallel (01\bar{1}0)_{Co}$.

The difference in Co structure (fcc and hcp structures for CoWB and CoWP thin films, respectively) is attributed to the difference in composition of the two films: the CoWB material is significantly richer in refractory metal than the CoWP alloy (see Table 5.5). Such a dependence of crystallographic structure on the composition of Co alloys has already been reported [EIN 05].

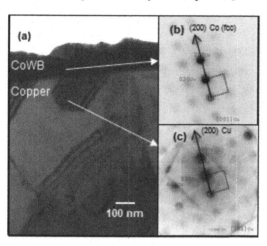

Figure 5.19. *a) TEM cross-section of a 150 nm-thick CoWB deposit and electron diffraction patterns of CoWB b) and copper c) across the interface showing alignment between the [100] direction of the fcc structures of both materials [OLI 10]*

For both materials, these results show that the orientation of the crystallized cobalt alloy is aligned with the underlying copper grains.

In the case of cubic CoWB, this can be understood by the limited lattice misfit with copper. The case of hexagonal CoWP is more complex due to the presence of the intermediate palladium layer. Although no direct observation of palladium crystal orientation can be made (this layer is much too thin), the heteroepitaxial growth of the cobalt alloy is probably the result of a heteroepitaxial growth of palladium on copper. Alternatively, the structure of the CoWP film could result from direct nucleation of the alloy on the copper surface in the vicinity of a palladium island. Indeed, the cathodic sites (Co deposition) may be remote from the anodic sites (oxidation of the reducing agent), because only the latter needs to be localized on the Pd islands.

Even though the mechanisms of heteroepitaxial growth of the cobalt alloys on copper are not fully understood, this particular behavior gives some arguments to explain the morphological variations of the films across copper grains.

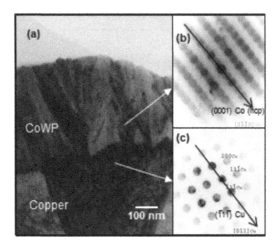

Figure 5.20. *a) TEM cross-section of a* 300 nm-*thick CoWP deposit and electron diffraction patterns of b) CoWP and c) copper across the interface showing alignment of the* [0001] *direction of the hcp structure of the CoWP with the* [Ī1Ī] *direction of the fcc structure of copper. Adapted From [OLI 10]*

Similarly to palladium deposition, specific interfacial affinities between the substrate and the deposit can modulate the energetic penalty associated with nucleation, leading to local variations of nucleation rate and density. Furthermore, differences in growth rate may exist between different crystallographic orientations. This could cause different growth rates between cobalt areas with different orientations and contribute to the final variations in the thickness of the deposit.

5.2.7. Undesired, parasitic nucleation of CoWP and CoWB

As explained in section 5.2.1, a central concern in the deposition of SAB is the *selectivity* of the process. But how can this property be quantified? A simple and efficient solution consists of testing the electrical performances of dedicated interconnect structures such as the interdigited comb/serpentines described in Figure 5.21. These structures are designed to increase the tiny current that always flows between neighboring lines through the (imperfect) dielectric material into a detectable range. Indeed, this arrangement maximizes the "proximity" length (at minimal spacing) between adjacent lines. Typically, this length can reach several meters.

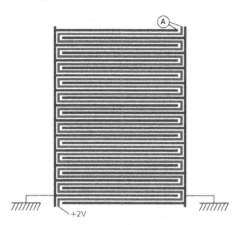

Figure 5.21. *Schematic representation of the interdigited combs and serpentine structure used to measure leakage current between adjacent conductors (line width:* 90 nm, *spacing:* 90 nm *or* 120 nm, *serpentine length:* 5.5 m)

5.2.7.1. *Evolution of leakage current with SAB thickness*

The evolution of the leakage current with cap thickness is plotted in Figure 5.22 for 90 nm/90 nm and 90 nm/120 nm comb/serpentines. This graph clearly shows two regimes. For both test structures, the median leakage current remains roughly constant, approximately one decade higher than the post-CMP reference, up to a critical thickness that depends on line spacing. This critical thickness is around 22 nm and 27 nm for 90 nm and 120 nm spacing, respectively. Above this critical thickness, the median leakage current dramatically increases.

5.2.7.2. *Leakage evolution below the critical thickness*

The top view SEM picture of a 90 nm/90 nm comb/serpentine structure covered with a 15 nm CoWP cap, i.e. below the critical thickness, is shown in Figure 5.23 (left). From the strict morphological standpoint, the CoWP deposit appears to be very selective on the copper lines with no visible defect on the dielectric surface. The CoWP deposit is also clearly located on top of Cu on the cross-section image in Figure 5.23 (middle). However, an electron energy loss spectroscopy (EELS) map shows that, in addition to the CoWP cap on top of the copper lines, a slight cobalt contamination is present at the surface of the interline dielectric. The +1 decade leakage current increase observed below the critical thickness in Figure 5.22 can reasonably be attributed to this contamination. The experimental evidence that leakage current is not proportional to the deposited CoWP thickness below the critical thickness is an indication that this residual contamination is independent of thickness and consequently probably does not consist of metallic cobalt but rather in cobalt ions trapped in the subsurface of the porous dielectric material, creating a conduction path as soon as voltage is applied to the structure.

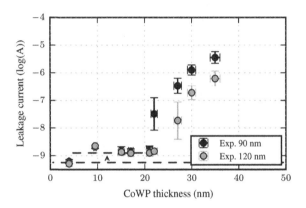

Figure 5.22. *Median leakage current (over* 30 *samples)*
in 90 nm/90 nm *and* 90 nm/120 nm *comb/serpentine structures as a*
function of CoWP cap thickness

5.2.7.3. *Leakage evolution above the critical thickness*

The comparison of top-view SEM pictures of 90 nm/120 nm comb/serpentines structures taken below and above the critical thickness are shown in Figure 5.24. They bring a first insight into the physical origin of leakage current increase above the critical thickness. The following two phenomena can be observed:

1) the presence of *parasitic CoWP nodules* randomly placed on the wafer surface with a size proportional to the cap thickness;

2) a lateral growth of the cap (due to the isotropic nature of the deposition reaction) leading to a progressive reduction of the minimal line spacing.

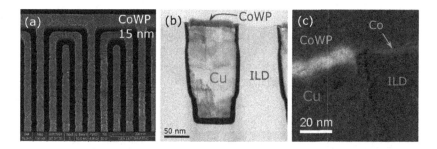

Figure 5.23. *Left: SEM picture of a* 90 nm/90 nm *comb/serpentine structure with a* 15 nm *CoWP cap. Middle: Corresponding TEM cross-section. Right: EELS map of cobalt showing a slight residual cobalt contamination on the interline dielectric surface. Adapted from [DEC 06]. For a color version of this figure, see www.iste.co.uk/haumesser/metals.zip*

Above a critical thickness for which the nodule diameter starts to exceed the remaining interline spacing, the combination of these two phenomena is likely to produce local short circuits between adjacent lines and cause the large increase in leakage currents. This analysis is consistent with the fact that the experimental critical thickness is larger for the 120 nm than for the 90 nm spacing.

The formation of these nodules is certainly the result of the *undesired nucleation of the Co-based alloy*. The location of this parasitic nucleation remains unclear, though. The presence of these defects on both the metal lines and the interline dielectric material suggests that they have not nucleated locally. Rather, they seem to have deposited there ready-formed (even if not fully grown). They have probably nucleated in the solution, or (more likely) on defects present at the surface of the dielectric material (such as metal residues after CMP). These weakly adherent nodules are lifted off the surface and redeposited at other places, such as the test structures studied here.

To quantify the impact of such defects on leakage current, a simple model can be established. This model is detailed in Figure 5.25. It accounts for the concomitant growth of the nodules and shrinking of interline spacing until the nodules bridge adjacent lines. Figure 5.26 shows that experimental leakage current results for 90 nm and 120 nm spacing are very well reproduced by this simple model using the fitting parameters R_c =1000 Ω (contact resistance between the nodules and the lines), A =1 nm^2 (section of the nodules), and $[Nod]$ =3 cm^{-2} (surface concentration of the nodules). In particular, the critical thickness above which leakage current increases dramatically is very well predicted for both spacing values. The evolution of the leakage amplitude above the critical thickness is also very well reproduced. The model, which predicts an abrupt increase in leakage at the thickness threshold, is in very good agreement with the much broader dispersion of the experimental data just above the threshold.

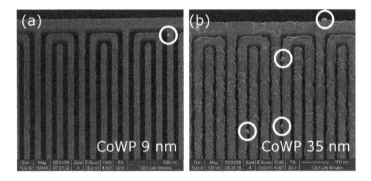

Figure 5.24. *SEM pictures of* 90 nm/120 nm *comb/serpentine structures with a CoWP cap of a)* 9 nm *(below critical thickness) and b)* 35 nm *(above critical thickness)*

The major conclusion that can be drawn from this model is that an extremely low density of parasitic nodules of $3\,\mathrm{cm}^{-2}$, *typically about ten per die, is responsible for an increase in leakage currents beyond acceptable values. Such nodules are most probably formed by parasitic nucleation, either in the solution or on impurities still present on the substrate.*

5.2.8. *Improvement of EM resistance*

Provided that the thickness of Co-based SABs remains below the critical value above which killer defects may be formed, the electroless processes are suited to form well-aligned metallic caps on top of the Cu lines. It is thus time to verify whether these deposits also bring the expected benefit of improved resistance toward EM.

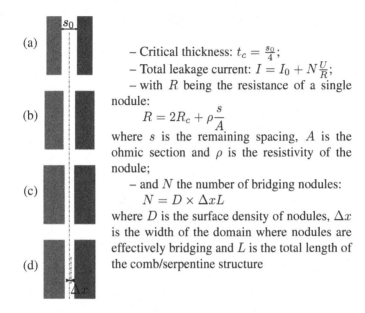

(a)

(b)

(c)

(d)

- Critical thickness: $t_c = \frac{s_0}{4}$;
- Total leakage current: $I = I_0 + N\frac{U}{R}$;
- with R being the resistance of a single nodule:
$$R = 2R_c + \rho\frac{s}{A}$$
where s is the remaining spacing, A is the ohmic section and ρ is the resistivity of the nodule;
- and N the number of bridging nodules:
$$N = D \times \Delta x L$$
where D is the surface density of nodules, Δx is the width of the domain where nodules are effectively bridging and L is the total length of the comb/serpentine structure

Figure 5.25. *Schematic representation of the origin of leakage currents: a) below critical thickness; b) at critical thickness; c) above critical thickness for a nodule centered between two lines; and d) above critical thickness for a non-centered nodule. Δx is the nodule center region where nodules are bridging adjacent lines. More details can be found in [OLI 08]*

For this purpose, specific test structures composed of $16\,120\,\mu m$ long lines connected to large pads can be used (Figure 5.27 (left))[4]. A certain amount of current passes through this structure under thermal stress and its resistance is monitored over time (Figure 5.27 (right)). During the test, the lines disrupt under the action of EM. As a result, incremental steps corresponding to the successive disruptions of individual lines are recorded in the electrical resistance (indicated by arrows in Figure 5.27 (right)). The resulting data are represented in a cumulative plot (Figure 5.28), allowing for determination of the MTTF.

Such experiments need to be reproduced under various temperatures and current densities (Table 5.6). The variation of MTTF with both factors gives access to the parameters of Black's equation. Interestingly enough, when a

---

4 Between these test lines, dummy lines are intercalated to simulate dense patterns.

12 nm-thick CoWP cap is deposited on the structures, *the activation energy deduced from these experiments is as high as* 1.8 eV. This is an important result, as it shows that the CoWP cap has significantly increased this activation energy (which is about 0.8–0.9 eV for conventional integration schemes, see section 5.2.1). Such an increase has been verified more thoroughly by other authors [HU 06]. This activation energy near 2 eV is usually associated with nucleation of voids in the bulk of Cu.

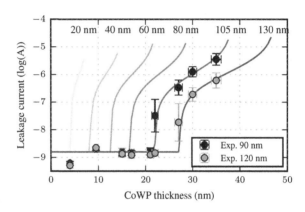

Figure 5.26. *Experimental and simulated median leakage currents as a function of CoWP cap thickness for various line spacings*

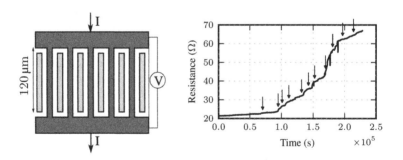

Figure 5.27. *Left: Test structure for EM measurements. Right: typical resistance evolution with time showing steps corresponding to individual line failure (arrows). Test conducted with a* 12 nm*-thick CoWP deposit, at* 256 °C, *under* 11.1 MA · cm^{-2}

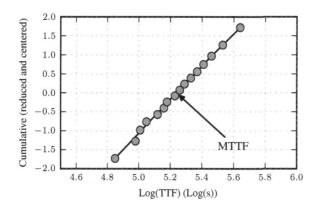

Figure 5.28. *Determination of the MTTF from the TTF values extracted from Figure 5.27. Experimental points are linearly fitted by the solid line. Its slope is $1/\sigma$ (σ =0.23). The MTTF is found to be 48.4 h (min=37 h, max=63 h)*

Test	Temperature (°C)	J (MA/cm2)	MTTF (h)	σ
1	255	24.0	24.3	0.21
2	316	29.1	0.4	0.20
3	256	11.1	48.4	0.23
4	316	31.7	0.4	0.15
E_a (eV)		1.8 ± 0.2		
n		1.0 ± 0.3		
σ		0.18 (min: 0.16, max: 0.24)		

Table 5.6. *Results of the EM tests: MTTF values measured at different temperatures and current densities and derived Black's parameters (see equation [5.2])*

A 12 nm *CoWP cap thus drastically increases the resistance of the structures toward EM. The activation energy associated with this phenomenon is doubled and reaches* 1.8 eV. *This corresponds to an extension of the lifetime under normal operating conditions by as much as* $\times 10^4$ *compared to the conventional capping technology.*

5.3. Conclusions

In this chapter, the importance of controlling nucleation processes has been demonstrated in the case of SLE and SAB deposition. In both cases, a sufficiently dense nucleation is needed to form thin and continuous metallic films. From the theoretical developments discussed in Chapters 2 and 3, the following aspects have been identified as critical to the success of both processes:

1) As both processes rely on heterogeneous nucleation mechanisms, surface preparation of the substrates is of paramount importance. In the case of SLE, it is necessary to remove the superficial oxide present on the nitride barrier layers, whereas electroless deposition requires extensive removal of corrosion inhibitors left by CMP, as well as appropriate activation of the Cu surface by Pd when hypophosphite is used as a reducing agent.

2) Beyond this, effects related to the crystallographic nature of the substrate may influence nucleation density, as was shown in the case of electroless deposition.

3) Electrolytic processes such as SLE, allow for extended control of the nucleation density through tuning of the deposition overpotential. This quantity is a direct measure of supersaturation.

Finally, the consideration of these aspects in the development of both processes allows for their successful integration. The SLE process efficiently repairs extremely thin and discontinuous PVD seed layers, as demonstrated by the electrical performances of the TSVs. Similarly, a spectacular enhancement of the resistance of damascene interconnect structures toward EM is obtained upon implementation of a CoWP SAB.

6

Nucleation and Stabilization of Metallic Nanoparticles in Ionic Liquids

Nucleation and growth of a solid (here a metal) from a liquid (here a solution, but this is also true for melts) is a natural bottom-up process by which isolated monomers may assemble into small objects. In Chapter 5, we showed that by controlling this precipitation, it is possible to tune and optimize the properties of thin metallic films. In this chapter, we shall extend this approach to the fabrication of smaller one-dimensional objects: *metallic nanoparticles (NPs)*. This subject was introduced in section 1.2. The main challenge resides in accurate control of the size of these objects at the nanometer scale. Also, these objects must remain stable over time. The general framework of NP synthesis through a chemical bottom-up approach was described in section 3.2. The role of stabilizing agents has been discussed to prevent aggregation of the NPs. Here, this chemical elaboration of metallic NPs is examined in further detail. As in the previous chapters, this challenge is approached through a specific example of current interest to the scientific community: the elaboration of metallic NPs in *ionic liquids (ILs)*.

6.1. What are ionic liquids?

The discovery of ILs is attributed to Walden in 1914. Walden described a salt, ethylammonium nitrate, that is liquid at room temperature. The field of ILs only become of general interest in the chemical community at the turn of this century. By definition[1], ILs are molten salts with a low melting point, near or even below room temperature. This is because they are composed of large organic cations, in which charge is delocalized, coupled with organic or inorganic anions. As a result, the coulombic interactions responsible for the cohesion of ionic materials are very weak in ILs compared to conventional salts. In turn, other interactions, such as hydrogen bonding, Van der Waals attraction or solvophobic interactions, may become significant. The chemical structure of ions suitable for the fabrication of ILs is highly diverse. There are several families of cations and anions, as depicted in Figure 6.1. In each family, the end groups borne by ions may be modulated both in their nature and structure (for instance alkyl chain length). As a result, the number of possible combinations is estimated to be as high as 10^{18}. Here, we will focus on one type of IL 1-alkyl-3-methylimidazolium bistrifluoromethyl-sulphonylimide, which combines imidazolium-based cations with sulfonylimide anions ($C_1 C_a ImNTf_2$, see Figure 6.2).

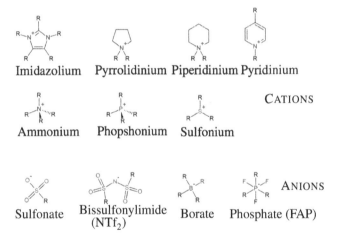

Imidazolium   Pyrrolidinium Piperidinium Pyridinium

Ammonium   Phopshonium   Sulfonium

CATIONS

Sulfonate   Bissulfonylimide (NTf$_2$)   Borate   Phosphate (FAP)

ANIONS

Figure 6.1. *Examples of ions*

In spite of this diversity in formulation, ILs share a unique combination of chemical and physical properties. They are very stable, virtually non-flammable, and can dissolve organic as well as inorganic species. They also are electrolytes with a wide electrochemical window (up to about 5 V) [END 02]. More surprisingly, they also exhibit extremely low vapor tensions

---

1 Usage dictates ILs be defined as salts having a melting point below $100\,^{\circ}C$.

comparable to solid salts: ILs easily sustain high vacuum conditions even at room temperature. This is particularly useful for practical purposes, direct TEM observation or X-ray photoelectron spectrometry (XPS) analysis, for instance. Another characteristic that these liquids share with most salts is their ability to capture water from the ambient atmosphere. Some of them are also sensitive to oxygen. To avoid contamination by water and air, ILs often need to be handled under controlled atmosphere (for instance in a glove box).

Finally, ILs exhibit some degree of self-organization [HAY 15, CHE 14, TRI 07]. A good example of such organization is given by the family of $C_1C_a ImNTf_2$ ILs. At the molecular level, the imidazolium cation and the anion form an ionic backbone, owing to the presence of a hydrogen bond between them. On both sides of this polar channel, the imidazolium ring bears two apolar alkyl chains: a methyl on one side and a longer alkyl chain (C_a, with $a = 2$ to 18) on the other. This molecular structure translates at a larger scale into a specific three-dimensional (3D) structure, which is accessible through numerical simulation [GUT 09]. In Figure 6.2, the structure of $C_1C_4 ImNTf_2$ is presented. In this structure, the apolar butyl chains are packed into small pockets about 1.4 nm in diameter, separated by a percolating polar network [PAD 07].

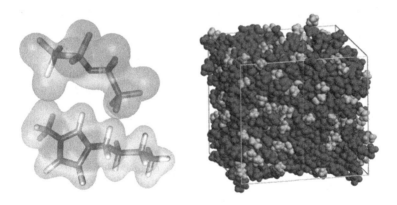

Figure 6.2. *Left: Developed formula of the 1-butyl-3-methylimidazolium bistrifluoromethyl-sulphonylimide ($C_1C_4 ImNTf_2$) ion pair. Right: 3D structure of $C_1C_4 ImNTf_2$. Adapted from [PAD 07]. For a color version of this figure, see www.iste.co.uk/haumesser/metals.zip*

In summary, the unusual properties of ILs result from a complex interplay between the chemical structure of their constituent ions and the various possible interactions between them. Now let us examine how these characteristics may be used for the synthesis and stabilization of metallic NPs.

6.2. Chemical synthesis of metallic NPs in ILs

The process by which suspensions of metallic NPs are synthesized is quite straightforward. It was inspired by syntheses first developed in common organic solvents [AYV 14, AMI 13]. The synthesis is carried out by decomposition of a suitable organometallic compound (OM) (for instance Ru(COD)(COT) to form Ru-NPs, see Figure 6.3 (left)) as follows [GUT 09, CAM 13]:

1) the precursor is dissolved in the IL (Figure 6.3 (middle));

2) this solution is brought to the desired temperature (typically 0 to 100 °C) in an autoclave under a vacuum;

3) the solution is then exposed to a pressure of H_2;

4) the autoclave is sealed and maintained at the same temperature for up to 72 h;

5) finally, the reaction is quenched by rapid cooling of the reactor to room temperature (if necessary) and evacuation of the gas phase.

All these steps are typically carried out in the strict absence of ambient air. By the end of this process, the solution has turned black (Figure 6.3 (right)), already a good indication that NPs have been formed. This can be supported by gas chromatography analysis of the gas phase which is necessary to quantify the volatile by-products of the decomposition reaction (cyclooctane in the case of Ru(COD)(COT)).

As mentioned in section 6.1, the suspensions can be readily observed under TEM. A typical image of a suspension of Ru-NPs is shown in Figure 6.4 which clearly shows that Ru has precipitated into small NPs around 2 nm in diameter. The corresponding size distribution is plotted in the same figure and fitted by a lognormal law, which affords the average diameter as well as size distribution of the NPs. Interestingly enough, this suspension is stable for at least several weeks (no significant evolution of the NP diameter is recorded) [CAM 10b].

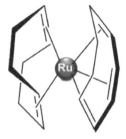  

Figure 6.3. *Left: Developed formula of the Ru(COD)(COT) precursor used in the synthesis of Ru-NPs. Middle: Ru(COD)(COT) solution in $C_1C_4ImNTf_2$. Right: Corresponding suspension of Ru-NPs after decomposition during* $40\,min$ *under* $0.3\,MPa$ *of H_2 at* $100\,°C$. *For a color version of this figure, see www.iste.co.uk/haumesser/metals.zip*

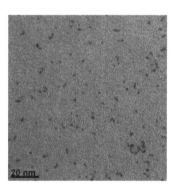 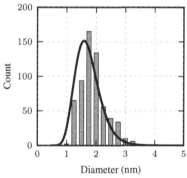

Figure 6.4. *Left: TEM image of Ru-NPs formed in $C_1C_8ImNTf_2$ at* $100\,°C$ *under* $0.4\,MPa$ H_2 *for* $4\,h$. *Right: Corresponding size distribution fitted by a lognormal law*

The simple decomposition under H_2 of solutions of an OM precursor dissolved in $C_1C_a ImNTf_2$ readily provides stable suspensions of metallic NPs with small size and narrow size distribution. No stabilizing agent is needed. Suitable precursors only need to:

– be soluble in the desired IL;

– decompose under H_2 into the zero-valent metal.

Several examples of metallic NPs successfully synthesized in these ILs are listed in Table 6.1. We are not aiming to comprehensively describe these results here. Nevertheless, we shall discuss two specific cases in the coming sections: Cu and Ta.

Metal	OM precursor	ILs Tested	Temperature	H_2 pressure	Duration	Diameter	Refs
Ru	Ru(COD) (COT)	$C_1C_a ImNTf_2$ (n=2–18)	0–100 °C	0.4–0.9 MPa	1–4 h	1.8–6 nm	[CAM 13]
Cu	CuMes	$C_1C_4 ImNTf_2$	100 °C	0.4–0.9 MPa	4 h	4–6 nm	[ARQ 12b]
Mn	MnNp$_2$	$C_1C_4 ImNTf_2$	100 °C	0.9 MPa	4 h	2.3 ± 0.5 nm	[ARQ 12a]
Ta	Np$_3$Ta= CHCMe$_3$	$C_1C_4 ImNTf_2$	25–100 °C	0.4 MPa	4–72 h	4–5 nm	[HEL 12]
Ni	Ni(COD)$_2$	$C_1C_6 ImNTf_2$	100 °C	0.9 MPa	4 h	5.3 ± 0.9 nm	

Table 6.1. *Precursors and conditions tested for the synthesis of metallic NPs in midazolium-based ILs*

6.2.1. *Synthesis of Cu-NPs*

The synthesis of Cu-NPs is carried out from mesitylcopper (CuMes, Figure 6.5 (left)) [ARQ 12b]. In this compound, Cu is in the +1 oxidation state. Therefore, its decomposition under H_2 involves the reduction of the metal, which requires additional energy compared to zero-valent precursors such as Ru(COD)(COT). For this reason, the decomposition of CuMes is performed at 100 °C. After 4 h under 0.9 MPa H_2 in $C_1C_4 ImNTf_2$, a well-dispersed suspension of NPs with a diameter of 5.0 ± 1.0 nm is obtained (Figure 6.5 (right)).

The high-resolution TEM image of one of these NPs clearly shows that they are crystalline (Figure 6.6 (left)). A period of 0.21 nm is visible in this image, which corresponds to the (111) distance in the fcc structure of Cu. Further evidence of their metallic nature is given by XPS coupled with Auger spectroscopy (Figure 6.6 (right)).

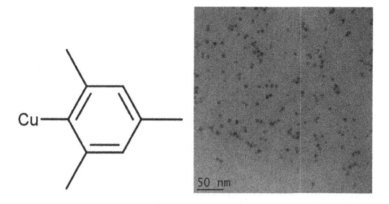

Figure 6.5. *Left: Developed formula of CuMes. Right: TEM image of Cu-NPs formed in $C_1C_4ImNTf_2$ at* $100\,°C$ *during* $4\,h$ *under* $0.9\,MPa\ H_2$

This example illustrates that well-crystallized NPs are obtained by this route. In the strict absence of ambient air, even oxygen-sensitive NPs such as Cu-NPs are obtained in the metallic state.

6.2.2. Synthesis of Ta-NPs

Tantalum is a metal of particular interest because of its very high melting point and excellent corrosion resistance, high thermal conductivity, small work function, plasticity and durability. Metallic tantalum has been used in electronics (see Chapters 1 and 5), the metallurgy industry, and in chemical and mechanical engineering [CAR 95]. The same properties for which Ta is an interesting metal make the synthesis of Ta-NPs difficult by conventional physical methods and the high oxophilicity of the metal means that forming and stabilizing Ta(0)-NPs is challenging [BAR 06].

Tris(neopentyl)neopentylidene tantalum(V), $Np_3Ta=CHCMe_3$, is a suitable precursor for the synthesis of Ta-NPs in ILs [HEL 12]. This OM compound is not soluble enough in $C_1CaImNTf_2$, though. To increase its solubility, pentane can be used as a co-solvent ($2\,wt\%$). Upon decomposition in $C_1C_4ImNTf_2$ under $0.4\,MPa\ H_2$ at $25\,°C$ for $72\,h$, a nicely stable suspension of NPs with a diameter of about $5\,nm$ is obtained. Again, the high resolution TEM analysis reveals the nicely crystalline structure of the NPs, with a period of $0.22\,\mathrm{nm}$ which is compatible with the (110) face of bulk fcc Ta (Figure 6.7).

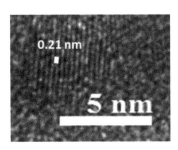 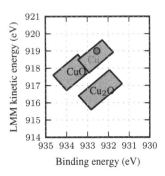

Figure 6.6. *Left: High-resolution TEM image image of an individual Cu-NP formed in $C_1C_4ImNTf_2$ at $100\,°C$ during $1\,h$ under $0.9\,MPa$ H_2 showing a crystalline structure compatible with metallic Cu. Adapted from [ARQ 12b]. Right: Wagner plot combining the XPS and Auger responses of the suspensions, confirming the metallic nature of the Cu-NPs. Adapted from [ARQ 13]*

This is a remarkable result, as this is probably the first demonstration of the synthesis of Ta-NPs at room temperature. Most processes described in the literature require much higher temperatures (typically $800\,°C$) [PAR 07, BAR 06].

In conclusion, ILs are well-suited media for the synthesis of metallic NPs under mild conditions [DUP 13]. In many instances, the NPs are well calibrated, with very narrow size distribution. An additional feature of many of these suspensions is their stability: many of them are stable for several weeks or longer. This is surprising, as usually metallic NPs coalesce under the action of attractive forces. As discussed in section 3.2.2, stabilizing agents are generally needed to prevent agglomeration. In ILs, no additive is present. This means that *ILs themselves are not innocent in the stabilization of the metallic NPs*. It is very likely that the ILs interact with the surface of the NPs to achieve this. Now, the question is: do these interactions significantly modify the energetic cost associated with the nucleation of the metallic NPs? In other words, do these interactions play a role in the final size of the metallic NPs? Are the Lamer and related mechanisms solely responsible for this final size, or do other phenomena specific to ILs contribute?

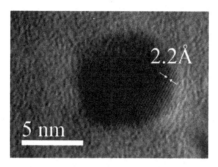

Figure 6.7. *High-resolution TEM image of a Ta-NP formed in* $C_1C_4ImNTf_2$ *under* $0.4\,\mathrm{MPa}\ H_2$ *at* $25\,°C$ *for* $72\,\mathrm{h}$

6.3. Interactions between ILs and NPs

The question of how ILs interact with solid surfaces in general and NPs in particular is not simple, nor is their subtle structuring in the bulk [HAY 15]. Again, depending on the chemical nature and structure of the ions, several phenomena may occur at the interface between ILs and solids. The possible mechanisms of interaction between ILs and NPs have recently been reviewed in detail [LUC 15]. In this section, we will focus on a shorter description of the main phenomena responsible for the stabilization of metallic NPs in ILs: viscous stabilization, which is a purely physical mechanism, and the formation of a protective IL layer around the NPs by electrostatic interactions (ion accumulation near the surface) or adsorption of ions (in the larger sense), which is more a chemical interaction.

6.3.1. *Viscous stabilization*

Rather than complex chemical effects, a simple physical reason may account for the enhanced stability of metallic NPs in ILs. When compared to most conventional solvents, ILs are rather viscous. Their viscosity varies considerably with temperature and alkyl chain length (Figure 6.8) [TOK 05]. More precisely, viscosity decreases with increasing temperature and decreasing alkyl chain length. Mass transport phenomena involved in the growth or coalescence processes are usually much less efficient in viscous media. This effect, referred to as *viscous stabilization*, cannot be excluded from contributing to the stability of metallic NPs in ILs.

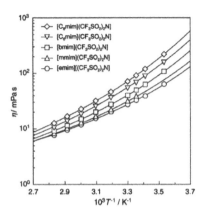

Figure 6.8. *Evolution of the viscosity of $C_1Calmntf_2$ with temperature and alkyl chain length [TOK 05]*

Qualitatively, metallic NPs should be more stable in more viscous ILs. For this reason, stability is supposed to increase with increasing alkyl chain length and decreasing temperature.

6.3.2. Electrostatic interactions

In ILs, the concept of EDL explained in section 3.2.2 still holds, but is (again) more complex than in diluted electrolytes. A major feature is the possible formation, near electrified surfaces, of multiple ionic layers alternating anions and cations (Figure 6.9). The organization of this interfacial layer of liquid is also impacted by other phenomena such as ion adsorption (see section 6.3.3), increasing the complexity of the problem. The interested reader is referred to recent in-depth reviews of the field [HAY 15, FED 14]. In ILs, we can expect that this dense EDL efficiently screens electrostatic repulsion [SZI 14].

According to the DLVO theory, we can thus expect that electrostatic phenomena do not significantly contribute to the stabilization of metallic NPs because they are screened by the concentrated ionic environment.

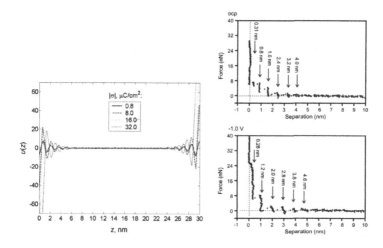

Figure 6.9. *Left: Simulated potential profiles for an IL between two oppositely charged electrodes with different charge densities [FED 08]. Right: Force versus distance profile measured by an AFM tip approaching an Au(111) surface in C_1C_4ImFAP at open circuit potential and $-1\,V$ versus Pt quasi-reference electrode [ATK 14]. For a color version of this figure, see www.iste.co.uk/haumesser/metals.zip*

6.3.3. Adsorption of ions

The most plausible mechanism to account for the stabilization of metallic NPs in ILs would be the formation of a protective ionic layer around the NPs similar to the mechanism of stabilization by coordinating agents in conventional solvents. There are several reasons why ions would stick to NPs and form such a protective layer [LUC 15].

A first possible mechanism would be the electrostatic adsorption of anions. Some authors have proposed that Ag-NPs are electrostatically stabilized by the preferential adsorption of anions in imidazolium-based ILs [RED 08]. This stabilization depends on the chemical nature and size of the anion.

In imidazolium-based ILs, the coordination of the metallic NPs by the imidazolium cation is supported by theoretical calculations. Numerical simulations performed with the system discussed in section 6.4 (i.e. Ru-NPs in $C_1C_4ImNTf_2$) have shown that imidazolium cations are adsorbed on the surface of Ru-NPs, with the alkyl chains preferentially directed away from the surface (Figure 6.10) [PEN 11]. Similar conclusions were drawn from surface-enhanced Raman spectroscopy experiments [SCH 07, RUB 08]. Such a coordination of the NPs by the imidazolium cations can be caused by electrostatic forces, but also by electronic interaction between the aromatic imidazolium ring and the NP. We cannot discard the possible stabilization by carbene species formed upon deprotonation of the cation [LUC 15]. Finally, we could also argue that this arrangement of the alkyl chains normal to the NP maximizes their apolar attraction (solvophobic interaction). After all, imidazolium cations, and especially those with long alkyl chains, are surfactants bearing a polar head (the imidazolium ring) and an apolar tail (the alkyl chain). In fact, similar ions have been used as surfactants in water to stabilize metallic NPs [DUP 13].

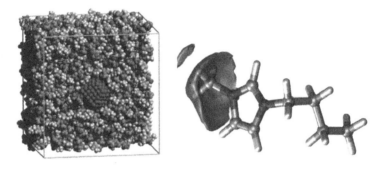

Figure 6.10. *Left: Simulation of a Ru-NP in $C_1C_4ImNTf_2$. Right: Spatial distribution function around the imidazolium ring of the cation showing the most probable locations of Ru atoms from the NP [PEN 11]. For a color version of this figure, see www.iste.co.uk/haumesser/metals.zip*

Other authors propose that in imidazolium-based ILs, the protective layer is not formed by one type of ion, but rather by polymeric entities composed of both anions and cations. The existence of such structures would result from the relatively strong hydrogen bonding between these ions [DUP 10]. These polymeric units could be cationic or anionic, depending on their relative content of cations and anions. It is suggested that, depending on their size, metallic NPs are either coordinated by anionic (smaller NPs) or by cationic species (larger NPs, Figure 6.11).

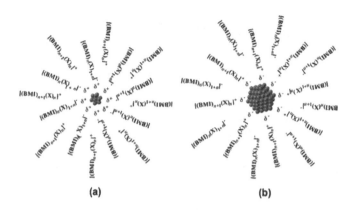

Figure 6.11. *Interaction with IL and resulting charge for a) small
(<10 nm) and b) large (>10 nm) NPs [DUP 10]*

*Experimental and theoretical evidence thus suggests that ions in ILs are
able to form a protective layer around the metallic NPs. This is the most
plausible mechanism to account for the exceptional ability of ILs (and
especially imidazolium-based ILs) to stabilize suspensions of metallic NPs.*

6.3.4. Impact of these interactions on nucleation, growth and aggregation

ILs are good media in which to prevent the aggregation of metallic NPs,
either by viscous stabilization or (more likely) by forming a coordinated
protecting layer around the NPs. However, these mechanisms may also
interfere in the classical nucleation and growth of these NPs. According to the
Lamer and associated models (see section 3.2.1), both nucleation and growth
contribute to the size of NPs. The temporal separation of both phases
(instantaneous nucleation) guarantees narrow size distribution. The
characteristics of the nucleation step (mainly the critical size of nuclei) fix the
number and average size of the NPs. Therefore, the possible impact of the
stabilizing mechanisms in ILs on both the nucleation and growth of the NPs
must be examined.

Both the viscous stabilization and the formation of a protective ionic layer
may retard or even block growth. High viscosities are usually associated with

low diffusivities. Hence, the supply of monomers to the growing supernuclei could be limited by diffusion in the viscous ILs. Also, the protective layer could act as a physical barrier to the incorporation of monomers.

In a situation where growth is sufficiently slow, the size of the NPs is directly determined by the number of supernuclei, thus the critical diameter of the nucleus. At a given supersaturation, the latter's size only depends on the energetic penalty $\phi(n)$ associated with the creation of the nucleus (see equation [2.8]). It is possible that this interfacial energy is affected by the structuration of the IL around the NP. Therefore, we cannot exclude that this phenomenon also affects nucleation and fixes the final size of the NPs.

It is the purpose of the follwoing section to examine these questions.

6.4. Size evolution of Ru-NPs

A general outcome of the theories of nucleation and growth described in Chapter 2 is that smallest objects are formed under conditions in which nucleation dominates. Qualitatively, the size of the clusters increases with the contribution of growth in the solidification. In this section, we examine how temperature and IL structure (alkyl chain borne by the imidazolium cation) may affect the final size of Ru-NPs. This will help us to estimate the effect of these conditions on the balance between nucleation and growth of Ru in these ILs.

The synthesis of Ru-NPs is schematically summarized in Figure 6.12. As mentioned in section 3.2.1, the decomposition of the OM precursor produces the monomers that undergo nucleation and growth to form the final NPs. In principle, this process may be described by the Lamer theory [LAM 50]. Following this reasoning, we can expect that:

1) if the nucleation and growth phases are well separated, well-calibrated NPs are obtained (instantaneous nucleation);

2) any acceleration of the decomposition reaction would lead to an enhanced supersaturation and higher nucleation rate (thus smaller NPs).

6.4.1. Conventional size control

A first interesting observation is that the size of Ru-NPs synthesized at $100\,°C$ in $C_1CaImNTf_2$ decreases as the length a of the alkyl chain on the cation increases (Figure 6.13). In $C_1C_4ImNTf_2$, $4\,nm$ NPs are formed. This diameter decreases to about $1.8\,\mathrm{nm}$ for $a = 8$ and remains stable for longer chains.

Figure 6.12. *Schematic representation of the nucleation and growth of Ru-NPs from Ru atoms produced by the decomposition of Ru(COD)(COT)*

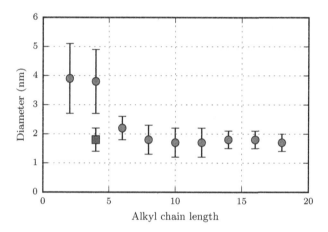

Figure 6.13. *Size evolution of Ru-NPs formed in $C_1CalmNTf_2$ at $100\,^{\circ}\mathrm{C}$ during $4\,\mathrm{h}$ under $0.9\,\mathrm{MPa}$ H_2. The square corresponds to the same experiment in $C_1C_4ImNTf_2$ with the addition of CuMes ($0.5\,\mathrm{at}\%$ as compared to the Ru content)*

This evolution can be rationalized in the conventional nucleation/growth description (Lamer model). We can expect that ILs with longer chains lead to enhanced stabilization of the NPs either by viscous mechanism (section 6.3.1) or specific coordination of the cation (section 6.3.3). As discussed in section 6.3.4, this would block the growth (and aggregation) of the NPs. Ultimately, the size of the final NPs would be that of the critical nuclei.

If we assume that this is the case[2], the observation that Ru-NPs have the same size in $C_1 Ca \mathrm{ImNTf}_2$ for $a = 8$ to 18 is very interesting. This would mean that the critical size of the nucleus – and thus the work of nucleation – does not depend on the molecular structure of the IL. As supersaturation is mainly driven by the rate of precursor decomposition (which is probably not affected by the structure of the cation), it should be the same in all these ILs. Therefore, the energetic penalty $\phi(n)$ associated with nucleation is not significantly modified in these various ILs (see equation [2.8]). We can conclude that the interaction of these ILs with the Ru-NPs does not significantly depend on the length of the alkyl chain.

By contrast, we would expect that a variation in temperature would affect NP size through modulation of the decomposition rate of Ru(COD)(COT). At higher temperature, faster decomposition is expected, leading to enhanced nucleation associated with the smaller critical size of the nucleus (and the reverse tendency at lower temperature). The size of Ru-NPs is found to decrease with increasing temperature in $C_1 C_{12} \mathrm{ImNTf}_2$ (Table 6.2). This observation thus corroborates our interpretation.

Temperature	Ru-NP diameter
75 °C	2.6 ± 0.3 nm
100 °C	1.8 ± 0.3 nm
150 °C	1.1 ± 0.2 nm

Table 6.2. *Size evolution of Ru-NPs formed in* $C_1 C_{12} ImNTf_2$ *during* 4 h *under* 0.9 MPa H_2.

Hence, it seems that nucleation alone is fixing the size of the NPs in long-chain ILs because in these compounds, growth is sufficiently suppressed. (This is not the case in ILs with short alkyl chains). Moreover, the nucleation itself (and thus the size of nuclei) is the same in all long-chain ILs. Their interaction with NPs thus seems to be quite independent of the alkyl chain, in line with the preferred coordination by the imidazolium ring. If this is true, we should be able to obtain small NPs (around 1.8 nm at 100 °C) in ILs with short chains by suppressing growth. In fact, this can be achieved by the incorporation in the reacting medium of an appropriate additive.

---

2 Ru-NPs of 1.8 nm in diameter contain about 150 atoms, which is a reasonable number for a critical nucleus.

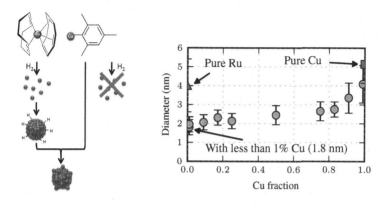

Figure 6.14. *Left: Mechanism of formation of Ru core-Cu shell bimetallic NPs. Right: Size evolution with Cu content of RuCu-NPs formed in $C_1C_4ImNTf_2$ at $100\,°C$ during $4\,h$ under $0.9\,MPa$ H_2. For a color version of this figure, see www.iste.co.uk/haumesser/metals.zip*

It turns out that upon mixing CuMes (the precursor to Cu-NPs described in section 6.2.1) with Ru(COD)(COT), smaller NPs are formed compared to the pure metals in $C_1C_4ImNTf_2$ at $100\,°C$ [HEL 13, ARQ 13]. As shown in Figure 6.13, Ru-NPs have a diameter of $3.9 \pm 0.9\,nm$. The size of Cu-NPs is $5 \pm 1\,nm$ (section 6.2.1). Upon mixing both precursors in a 1:1 ratio (keeping the total metal concentration constant), NPs as small as $2.5 \pm 0.6\,nm$ are formed (Figure 6.14). These NPs have been extensively characterized [HEL 14]. They exhibit a core–shell structure, with Ru in the core and Cu in the shell. This particular structure has been assigned to the difference in decomposition kinetics of the two precursors. Ru(COD)(COT) is decomposed faster and forms Ru-NPs on which CuMes reacts. It seems that this reaction inhibits further growth of Ru[3], which is the only explanation to account for smaller bimetallic NPs. A very important observation for our purpose here is that this mechanism is active for extremely low amounts of CuMes (as low as 0.5%). Upon addition of this small amount of CuMes, NPs as small as $1.8 \pm 0.5\,nm$ are obtained. Hence, *CuMes can be considered as an additive that suppresses growth of Ru-NPs*. Interestingly enough, when this additive is present in $C_1C_4ImNTf_2$, the size of the NPs is the same as for long-chain ILs (see the red square in Figure 6.13).

---

3 Or aggregative growth of the Ru-NPs?

At $100\,°C$, *the evolution in size of Ru-NPs reflects the balance between nucleation and growth, according to the conventional nucleation theories in continuous media. More precisely, the process can be described by the Lamer model, in which nucleation and growth phases are well separated (instantaneous nucleation). This accounts for the small size distribution of the Ru-NPs. Moreover, the following conclusions can be drawn:*

– for ILs with a long alkyl chain, the growth mechanism is blocked, and the size of Ru-NPs is fixed by nucleation (i.e. it corresponds to the critical size for nucleation);

– the interaction responsible for blocking growth is the same in all ILs with long chains, resulting in similar nucleation kinetics and NP size;

– growth can also be suppressed in short-chain ILs by adding small amounts of CuMes. In this case, small Ru-NPs at (or near to) the critical size are obtained;

– this size only depends on temperature, as expected: the higher the temperature, the faster the nucleation, the smaller this size.

6.4.2. *Unconventional size control*

Both the low melting point of the short-chain ILs and the reactivity of Ru(COD)(COT) allow the reaction to be performed at a much lower temperature. In Figure 6.15, the evolution in size of Ru-NPs in $C_1C_4ImNTf_2$ is shown when temperature is decreased from 75 to $0\,°C$. According to the discussion in section 6.4.1, we would expect the size to increase at lower temperatures. However, the opposite tendency is observed: the diameter decreases from $3.1 \pm 0.7\,nm$ at $75\,°C$ to $1.1 \pm 0.2\,nm$ at $0\,°C$. This tendency and the very small diameter of the NPs at $0\,°C$ clearly cannot be accounted for by the same model as discussed in section 6.4.1.

Furthermore, the size evolution with alkyl chain length, displayed in Figure 6.16, is significantly different from the variation observed at $100\,°C$. In this case, this evolution is not monotonic [GUT 09]. In $C_1C_2ImNTf_2$, Ru-NPs of 2.3 ± 0.6 are obtained, which are larger than those in $C_1C_4ImNTf_2$ $(1.1 \pm 0.2\,nm)$. So the size decreases drastically between $a = 2$ and $a = 4$. As the alkyl chain length increases to $a = 6$ and $a = 8$, so does the diameter of RuNPs.

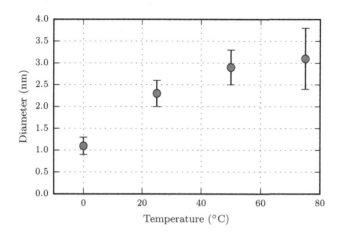

Figure 6.15. *Size evolution of Ru-NPs formed in $C_1C_4ImNTf_2$ at various temperatures over* 72 h *[CAM 10a]*

The Ru-NPs in $C_1C_4ImNTf_2$ are thus exceptionally small when compared to the other ILs. Moreover, the increase in size for longer-chain ILs is opposite to the tendency recorded at 100 °C. Therefore, an additional phenomenon must be involved in the low-temperature synthesis of Ru-NPs.

As explained in section 6.1, ILs possess an internal organization at the nanometer scale [HAY 15]. Our imidazolium-based ILs, in particular, are characterized by a separation between polar and apolar moieties (see Figure 6.2). In $C_1C_2ImNTf_2$, this separation has no significant consequence: the short methyl and ethyl groups are simply "dissolved" in the polar liquid, and this IL may be considered a continuous medium. In $C_1C_4ImNTf_2$, the situation is different. The butyl chains have enough volume to form isolated apolar pockets, as shown in Figure 6.2. Interestingly enough, the size of these pockets is about 1.3 nm [TRI 07]. Hence, the size of Ru-NPs in this IL corresponds to the size of the apolar pockets. Even more interestingly, this observation holds for $C_1C_6ImNTf_2$ and $C_1C_8ImNTf_2$. As the alkyl chain length increases, so does the size of apolar pockets[4]. Again, there is a good correlation with the size of Ru-NPs (see Figure 6.16).

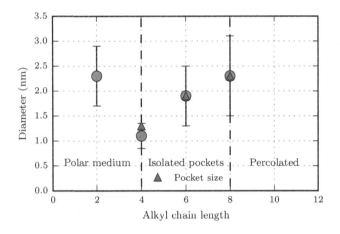

Figure 6.16. *Size evolution of Ru-NPs formed at* $0\,^\circ C$ *in* $C_1CalmNTf_2$. *The size of the NPs (circles) is compared with the size of the apolar pockets in the IL (triangles, also see Figure 6.2) [GUT 09]*

From these observations, the unusual size evolution of Ru-NPs in $C_1CalmNTf_2$ ($a =2$–8) at $0\,^\circ C$ can be attributed to the three-dimensional structure of the ILs. It is reasonable to suppose that Ru(COD)(COT) preferentially dissolves in the organic-like alkyl part of the medium. Then, the nucleation of Ru occurs in the apolar pockets, and the growth of Ru-NPs is constrained by their limited size.

At higher temperatures or for the shortest chains ($a = 2$), this granularity of the ILs would either be lost or no longer constrain the distribution of Ru(COD)(COT). Thus, they would behave like continuous media, which is

---

4 This rules out viscous stabilization mechanisms accounting for smaller NPs at lower temperature. Viscosity is also expected to increase with alkyl chain length (see Figure 6.8).

one basic assumption that is required to apply classical nucleation and growth models (see section 6.4.1).

6.5. Conclusions

Due to their unique and tunable properties, metallic NPs offer many opportunities for current and future technological development. However, to be usable in actual applications, they must be fabricated through manufacturable and reliable processes. These processes must yield NPs with calibrated size. Chemical routes in which metallic NPs are formed from the bottom-up assembly of isolated atoms in solution are promising in this regard, provided that:

1) the nucleation step is short and separated from the subsequent growth (instantaneous nucleation);

2) an efficient mechanism is found to stabilize the NPs and limit their size.

For this purpose, conventional chemical processes require the addition of specific stabilizing agents that coordinate the NPs and block further growth or coalescence. Here we have introduced an original approach in which such additives are not needed. This process relies on the unusual properties of ILs. These media are shown to act as "smart" solvents for the synthesis of metallic NPs by a simple process in which an OM precursor is decomposed under H_2 to precipitate metallic NPs. Interestingly enough, the IL itself acts as a stabilizer because it forms a protective layer around the NPs. As a result, NPs with controllable size can be obtained in these media according to conventional nucleation and growth mechanisms (Lamer model).

In addition, the self-organization of ILs at the nanometer scale further control the size. It is shown that under appropriate conditions, the IL acts as a template that constrains the size of the NPs. This effect is particularly efficient at forming extremely small NPs (around 1 nm in diameter) with extremely narrow size distribution. Moreover, playing with the structure of the IL, this size can be tuned over quite a large range.

Appendices

Appendix 1

Demonstration of the General Expression for the Nucleation Rate

Here we develop the reasoning leading to the establishment of equation [2.34].

Equation [2.33] can be rewritten as:

$$J(t) = \lim_{\Delta t \to 0} \frac{\left(\sum_{n^*(t+\Delta t)+1}^{M(t+\Delta t)} Z_n(t + \Delta t)\right) - \left(\sum_{n^*(t)+1}^{M(t)} Z_n(t)\right)}{\Delta t} \qquad [A1.1]$$

Let us assume that:

– during interval Δt, the variation of n^*, Δn^*, is finite, i.e. $n^*(t + \Delta t) = n^*(t) + \Delta n^*$;

– $\forall m \geq \min(M(t), M(t + \Delta t))$, $Z_m(t) = 0$;

– $\forall t \in [t; \Delta t]$, $\forall m \in [n^*(t); n^*(t) + \Delta n^*]$, $Z_m(t) \simeq Z_{n^*}(t)$.

Then

$$\left(\sum_{n^*(t+\Delta t)+1}^{M(t+\Delta t)} Z_n(t + \Delta t)\right) \simeq \sum_{n^*(t)+1}^{M(t)} Z_n(t + \Delta t) - \Delta n^* Z_{n^*}(t) \qquad [A1.2]$$

and

$$J(t) = \lim_{\Delta t \to 0} \left[\sum_{n^*(t)+1}^{M(t)} \frac{Z_n(t + \Delta t) - Z_n(t)}{\Delta t} - \frac{\Delta n^*}{\Delta t} Z_{n^*}(t) \right] \qquad [A1.3]$$

$$J(t) = \sum_{n^*(t)+1}^{M(t)} \frac{dZ_n(t)}{dt} - \frac{dn^*}{dt} Z_{n^*}(t) \qquad [A1.4]$$

Owing to equation [2.110]:

$$J(t) = j^*(t) - j_{M(t)}(t) + \sum_{n^*(t)+1}^{M(t)} (K_n(t) - L_n(t)) - \frac{dn^*}{dt} Z_{n^*}(t) \qquad [A1.5]$$

As $j_{M(t)}(t) = 0$ and considering the case where supernuclei are not created or consumed by other processes than attachment and detachment of monomers (i.e. $K_n(t) - L_n(t) = 0 \; \forall m \geq n^*(t)$), this results in the following:

$$J(t) = j^*(t) - \frac{dn^*}{dt} Z^*(t) \qquad [A1.6]$$

Appendix 2

Stationary Cluster Population and Nucleation Rate

A2.1. For discrete n values

The purpose is to solve the system of $M - 1$ equations in condition 3 (page 26):

$$\forall n \in [1; M - 1], j_n = F_n Z_n - B_{n+1} Z_{n+1} = J_s = \text{constant}$$

Considering the boundary condition equation [2.36], the last two equations of this system pass into:

$$J_s = F_{M-1} Z_{M-1} = F_{M-2} Z_{M-2} - B_{M-1} Z_{M-1} \qquad [A2.1]$$

Considering equation [2.13] and replacing B_{M-1}, Z_{M-2} can be expressed as:

$$Z_{M-2} = C_{M-2}(F_{M-1} Z_{M-1}) \left(\frac{1}{F_{M-2} C_{M-2}} + \frac{1}{F_{M-1} C_{M-1}} \right) \qquad [A2.2]$$

Based on the result, let us assume that:

$$Z_n = C_n(F_{M-1} Z_{M-1}) \left(\sum_{m=n}^{M-1} \frac{1}{F_m C_m} \right) \qquad [A2.3]$$

Then:

$$F_{n-1}Z_{n-1} - B_n Z_n = J_s = F_{M-1}Z_{M-1} \qquad [A2.4]$$

$$F_{n-1}Z_{n-1} - \frac{C_{n-1}}{C_n}Z_n = J_s = F_{M-1}Z_{M-1} \qquad [A2.5]$$

which yields

$$Z_{n-1} = C_{n-1}(F_{M-1}Z_{M-1}) \left(\frac{1}{F_{n-1}C_{n-1}} + \sum_{m=n}^{M-1} \frac{1}{F_m C_m} \right) \quad [A2.6]$$

which shows that equation [A2.3] is the correct definition of Z_n. Finally, combining this relation at $n = 1$ with the boundary condition $Z_1 = C_1$ (equation [2.37]):

$$(F_{M-1}Z_{M-1}) \left(\sum_{m=1}^{M-1} \frac{1}{F_m C_m} \right) = 1 \qquad [A2.7]$$

allows for replacing $F_{M-1}Z_{M-1}$ in equations [A2.1] and [A2.3] to yield equations [2.39] and [2.38].

A2.2. For continuous n values

In this case, the demonstration is more straightforward, as the combination of equation [2.41] and condition 3 (equation [2.40]) leads to the differential equation:

$$\frac{\partial Z(n,t)}{\partial t} = \frac{d}{dn}\left[F(n)C(n)\frac{d}{dn}\left(\frac{Z(n)}{C(n)} \right) \right] = 0 \qquad [A2.8]$$

This easily leads to

$$F(n)C(n)\frac{d}{dn}\left(\frac{Z(n)}{C(n)} \right) = A \qquad [A2.9]$$

where A is a constant. This new differential equation is also readily resolved into:

$$\frac{Z(n)}{C(n)} = B + \int_1^n \frac{A}{F(m)C(m)}\,\mathrm{dm} \qquad\text{[A2.10]}$$

where B is a constant. A and B are determined from the boundary conditions according to:

$$\frac{Z(1)}{C(1)} = 1 = B \qquad\text{[A2.11]}$$

$$Z(M) = 0 = 1 + A\int_1^M \frac{1}{F(m)C(m)}\,\mathrm{dm} \qquad\text{[A2.12]}$$

readily leading to equations [2.77] and [2.78].

Appendix 3

Where is the Surface of the Cluster?

This question proceeds from a more general concern in the treatment of interfaces in thermodynamics. Indeed, in actual systems, an interface is never perfectly abrupt but corresponds to a continuous variation of the density (or density gradient) from one phase to the other on a short, but finite distance (Figure A3.1). Since an exact description of such a situation would be overcomplicated, two simplified approaches are generally considered.

In the first approach, developed by Guggenheim [GUG 85], the interfacial region is treated as a separate phase. This phase is narrow but needs to extend in both main phases (usually a few nanometers), so its boundaries are in regions where these phases have their nominal composition (Figure A3.1). Therefore, this formalism is not well suited to describe the interface of small objects such as clusters whose size becomes comparable to the extension of this interfacial region.

For this reason, this question is rather approached following Gibbs' formalism [KAS 00]. In this case, the interface is described as a mathematical surface of zero thickness. Consequently, the interface is perfectly abrupt. To account for the existence of the composition gradient in the interfacial region, a specific composition needs to be attributed to this surface. Such a situation is depicted in Figure A3.1 in the case of the nucleation of a condensed phase, in which the cluster is a region of higher concentration of species (this figure can easily be adapted to the opposite case).

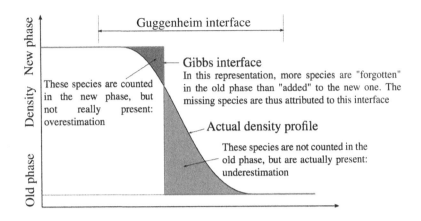

Figure A3.1. *Gibbs' description of an interface (and comparison with Guggenheim's formalism)*

Once the interface is placed, the volume of the cluster (V_{cl}) is defined, separating the species within the cluster $(n = c_{cl}V_{cl})$ from the species still in the old phase $(c_{old}(V - V_{cl})$ such species). If the interface is placed within the gradient region, the number of species in the cluster is overestimated, and the number of species in the old phase is underestimated. For instance, if the interface is "closer" to the cluster (as depicted in Figure A3.1), the number of species within the cluster is only slightly overestimated, whereas the number of species in the old phase is largely underestimated. As a result, the sum of the two populations does not correspond to the total number of species in the system: $c_{cl}V_{cl} + c_{old}(V - V_{cl}) < M$. To obey mass balance, the n_i "missing" species ($n_i > 0$ in this case) are thus attributed to the interface. Conversely, if the interface is placed "away" from the cluster, $n_i < 0$. In addition to mass balance, energy balance must be respected as well. Therefore, the appropriate chemical potential μ_i is attributed to each of these n_i species.

By definition, the interface defined by $n_i = 0$ is called equimolar. This choice of interface usually leads to a simpler formalism within the classical theory of nucleation.

Glossary

3D integration Approach in which several layers of devices are stacked to increase their number per unit surface area of substrate.

Accelerator Chemical compound that can complex Cu^+ ions and adsorb on the Cu surface of the substrate. This adsorbate facilitates charge transfer during electrodeposition, counteracting the action of the suppressor.

Adatom Neutral atom adsorbed on a substrate, seeking for a stable crystallographic site (either by incorporation into an existing crystal lattice or by nucleation).

Cluster Small grain of the new phase. Its size is usually defined by the number n of its contituting monomers. As such, a cluster may be considered as an n-mer of the new phase.

Conformal deposition Growth regime in which deposited thickness does not depend on local topography. During filling of a hollow structure, the growth rate is the same everywhere within and out of the structure.

CoWB Electroless process for the deposition of a Co-based self-aligned barrier, which uses morpholine borane as a reducing agent, and readily metallises Cu surfaces. A tungsten salt is also present in the solution, leading to the co-deposition of W with Co.

CoWP Electroless process for the deposition of a Co-based self-aligned barrier, which uses hypophosphite as a reducing agent, and requires activation of Cu surface by Pd prior to deposition. A tungsten salt is also present in the solution, leading to the co-deposition of W with Co.

Damascene Process in which metal lines are formed by filling hollow structures etched in a dielectric material.

Electroless deposition Process by which metallic ions are reduced into the metal under the action of a reducing agent. This reaction is usually kinetically limited to catalytically active surfaces, the deposited metal being a catalyst for the reaction (*autocatalytic* process).

Extreme fill Extreme case of superconformal deposition, in which the differential of growth rate between inhibited (field) and non-inhibited areas (via bottom) exceeds 50x. This regime is necessary to achieve void-free fill of TSVs.

Inhibitor Chemical compound that inhibits electrolytic deposition of Cu in TSV fill processes. By contrast with a suppressor used for damascene applications, this additive cannot be desorbed from the metal surface.

Monomer Unit element of the new phase. A monomer can be an atom, a molecule, etc. By definition, an n-sized cluster of the new phase contains n monomers.

Nucleus Cluster at the critical size.

Overpotential Difference between the potential applied to an electrode and its equilbrium potential.

Seed layer Thin conductive film on which an electroplating process can be initiated.

Subconformal deposition Growth regime during filling of a hollow structure, in which the growth rate on the field of the substrate exceeds the growth rate within the structure.

Subnucleus Cluster below the critical size.

Superconformal deposition Growth regime during filling of a hollow structure, in which the growth rate at the bottom exceeds the growth rate on the field of the substrate.

Supernucleus Cluster above the critical size.

Supersaturation Favorable balance in Gibbs energy which drives the phase transformation. The larger this difference (in absolute value), the more favorable will be the phase transformation.

Suppressor Chemical compound that inhibits electrolytic deposition of Cu. It is usually a polymer, which adsorbs on the Cu surface of the substrate. This adsorbate blocks the surface from the solution and prevents discharge of Cu^{2+} ions.

List of Acronyms

AFM atomic force microscopy.

AR aspect ratio between the depth and the width of a hollow structure.

ASAXS anomalous small-angle X-ray scattering.

CE counter electrode.

CEAC model to explain the accumulation of accelerator at the bottom of damascene trenches and vias during Cu filling.

CMP chemical mechanical polishing.

CVD chemical vapor deposition.

DLS dynamic light scattering.

DLVO Derjaguin–Landau–Verwey–Overbeek theory.

DMAB $(CH_3)_2NHBH_3$.

DoB process of Cu electroplating onto resistive barrier layers.

EBSD electron backscatter diffraction.

EDL electrical double layer.

EDX energy-dispersive X-ray spectroscopy.

EELS electron energy loss spectroscopy.

EM electromigration, thermally activated drift of metal atoms under the flow of a high electrical current.

EN ethylenediamine.

fcc face-centered cubic crystalline structure.

FIB focused ion beam.

GC gas chromatography.

hcp hexagonal close-packed crystalline structure.

HRTEM high resolution transmission electron microscopy.

IC integrated circuit.

IL ionic liquid.

JMAK Johnson-Mehl-Avrami-Kolmogorov equation.

LA low acid Cu electrolytes contain about 0.1mol/L sulfuric acid.

MA medium acid Cu electrolytes contain about 0.5mol/L sulfuric acid.

MPS mercaptopropane sulfonic acid.

MTTF mean time to failure.

NP nanoparticle whose size is typically below 100 nm.

NRA nuclear reaction analysis.

OM organometallic compound.

PECVD plasma-enhanced chemical vapor deposition.

PIXE particle induced x-ray emission.

PVD physical vapor deposition.

RBS rutherford back-scattering spectroscopy.

RDE rotating disk electrode.

RE reference electrode.

SAB self-aligned barrier.

SCE saturated calomel electrode.

SEM scanning electron microscopy.

SERS surface-enhanced raman spectroscopy.

SHE standard hydrogen electrode.

SIMS secondary ion mass spectroscopy.

SLE process of Cu electroplating aiming at bridging discontinuities in a PVD seed layer.

SPS bis(sodiumsulfopropyl)disulfide.

TEM transmission electron microscopy.

TMAH tetramethylhydroxide.

TSV deep contact holes to connect stacked layers in the 3D integration.

TTF time to failure.

WE working electrode.

XPS X-ray photoelectron spectrometry.

XRF X-ray fluorescence.

XRR X-ray reflectometry.

Bibliography

[ALM 07] ALMOG R., SVERDLOV Y., GOLDFARB I. *et al.*, "CoWBP capping barrier layer for sub 90 nm Cu interconnects", *Microelectronic Engineering*, vol. 84, no. 11, pp. 2450–2454, 2007.

[AMI 13] AMIENS C., CHAUDRET B., CIUCULESCU-PRADINES D. *et al.*, "Organometallic approach for the synthesis of nanostructures", *New Journal of Chemistry*, vol. 37, pp. 3374–3401, 2013.

[AND 98] ANDRICACOS P., UZOH C., DUKOVIC J. *et al.*, "Damascene copper electroplating for chip interconnections", *IBM Journal of Research and Development*, vol. 42, no. 5, pp. 567–574, 1998.

[ANT 06] ANTONELLI S., ALLEN T., JOHNSON D. *et al.*, "Determining the role of W in suppressing crystallization of electroless Ni-W-P films", *Journal of the Electrochemical Society*, vol. 153, no. 6, pp. J46–J49, 2006.

[ARM 11] ARMINI S., TOKEI Z., VOLDERS H. *et al.*, "Impact of terminal effect on Cu electrochemical deposition: filling capability for different metallization options", *Microelectronic Engineering*, vol. 88, no. 5, pp. 754–759, 2011.

[ARQ 12a] ARQUILLIÈRE P., Nanoparticules mono- et bimétalliques pour la métallisation de microvias par un procédé innovant utilisant les liquides ioniques, PhD Thesis, University of Lyon 1, 2012.

[ARQ 12b] ARQUILLIÈRE P., HAUMESSER P., SANTINI C., "Copper nanoparticles generated in situ in imidazolium based ionic liquids", *Microelectronic Engineering*, vol. 92, pp. 149–151, 2012.

[ARQ 13] ARQUILLIÈRE P., HELGADOTTIR I.S., SANTINI C.C. *et al.*, "Bimetallic Ru–Cu nanoparticles synthesized in ionic liquids: kinetically controlled size and structure", *Topics in Catalysis*, vol. 56, pp. 1192–1198, 2013.

[ATK 14] ATKIN R., BORISENKO N., DRÜSCHLER M. *et al.*, "Structure and dynamics of the interfacial layer between ionic liquids and electrode materials", *Journal of Molecular Liquids*, vol. 192, pp. 44–54, 2014.

[AYV 14] AYVALI T., LECANTE P., FAZZINI P.-F. *et al.*, "Facile synthesis of ultra-small rhenium nanoparticles", *Chemical Communications*, vol. 50, pp. 10809–10811, 2014.

[BAL 05] BALZANI V., "Nanoscience and nanotechnology: a personal view of a chemist", *Small*, vol. 1, no. 3, pp. 278–283, 2005.

[BAR 48] BARDEEN J., BRATTAIN W.H., "The transistor, a semi-conductor triode", *Physical Review*, vol. 74, pp. 230–231, July 1948.

[BAR 06] BARR J., AXELBAUM R., MACIAS M., "Processing salt-encapsulated tantalum nanoparticles for high purity, ultra high surface area applications", *Journal of Nanoparticle Research*, vol. 8, no. 1, pp. 11–22, 2006.

[BAR 08] BARRIÈRE C., ALCARAZ G., MARGEAT O. *et al.*, "Copper nanoparticles and organometallic chemical liquid deposition (OMCLD) for substrate metallization", *Journal of Materials Chemistry*, vol. 18, no. 26, pp. 3084–3086, 2008.

[BAU 58] BAUER E., "Phänomenologische Theorie der Kristallabscheidung an Oberflächen. I", *Zeitschrift für Kristallographie – Crystalline Materials*, vol. 110, p. 372, 1958.

[BUD 00] BUDEVSKI E., STAIKOV G., LORENZ W., "Electrocrystallization: nucleation and growth phenomena", *Electrochimica Acta*, vol. 45, no. 15, pp. 2559–2574, 2000.

[CAM 10a] CAMPBELL P., SANTINI C., BAYARD F. *et al.*, "Olefin hydrogenation by ruthenium nanoparticles in ionic liquid media: does size matter?", *Journal of Catalysis*, vol. 275, no. 1, pp. 99–107, 2010.

[CAM 10b] CAMPBELL P., SANTINI C., BOUCHU D. *et al.*, "A novel stabilisation model for ruthenium nanoparticles in imidazolium ionic liquids: in situ spectroscopic and labelling evidence", *Physical Chemistry Chemical Physics*, vol. 12, pp. 4217–4223 2010.

[CAM 13] CAMPBELL P., PRECHTL M., SANTINI C. *et al.*, "Ruthenium nanoparticles in ionic liquids a saga", *Current Organic Chemistry*, vol. 17, no. 4, pp. 414–429, 2013.

[CAR 95] CARDONNE S., KUMAR P., MICHALUK C. *et al.*, "Tantalum and its alloys", *International Journal of Refractory Metals and Hard Materials*, vol. 13, no. 4, pp. 187–194, 1995.

[CAY 07] CAYRON C., "Multiple twinning in cubic crystals: geometric/algebraic study and its application for the identification of the $\Sigma 3^n$ grain boundaries", *Acta Crystallographica Section A*, vol. 63, no. 1, pp. 11–29, January 2007.

[CHA 07] CHARLES-ALFRED C., JOUSSEAUME V., "a-Sic:H low-k deposition as copper diffusion barrier layer in advanced microelectronic interconnections", *Surface and Coatings Technology*, vol. 201, nos. 22–23, pp. 9260–9263, 2007.

[CHE 14] CHEN S., ZHANG S., LIU X. *et al.*, "Ionic liquid clusters: structure, formation mechanism, and effect on the behavior of ionic liquids", *Physical Chemistry Chemical Physics*, vol. 16, pp. 5893–5906, 2014.

[CHO 08] CHOU Y., SUNG Y., BAI C. *et al.*, "Effects of molybdate concentration on the characteristics of Ni-Mo-P diffusion barriers grown by nonisothermal electroless deposition", *Journal of the Electrochemical Society*, vol. 155, pp. D551–D557, 2008.

[COR 13] CORMARY B., DUMESTRE F., LIAKAKOS N. *et al.*, "Organometallic precursors of nano-objects, a critical view", *Dalton Transactions*, vol. 42, pp. 12546–12553, 2013.

[CUZ 10] CUZZOCRÉA J., DERONZIER E., HAUMESSER P. *et al.*, "Seed layer enhancement: an efficient process for the fabrication of 3D interconnects", *Advanced Metallization Conference (AMC)*, pp. 253–258, 2010.

[DA 05] DA SILVA S., MOURIER T., HAUMESSER P. *et al.*, "Gap fill enhancement with medium acid electrolyte for the 45 nm node and below", *Advanced Metallization Conference (AMC 2005)*, pp. 513–517, 2005.

[DE 05] DE SIERVO A., PANIAGO R., SOARES E. *et al.*, "Growth study of Cu/Pd(111) by RHEED and XPS", *Surface Science*, vol. 575, nos. 1–2, pp. 217–222, 2005.

[DEC 06] DECORPS T., HAUMESSER P., OLIVIER S. *et al.*, "Electroless deposition of CoWP: material characterization and process optimization on 300 mm wafers", *Microelectronic Engineering*, vol. 83, no. 11–12, pp. 2082–2087, 2006.

[DIJ 10] DIJON J., SZKUTNIK P., FOURNIER A. *et al.*, "How to switch from a tip to base growth mechanism in carbon nanotube growth by catalytic chemical vapour deposition", *Carbon*, vol. 48, no. 13, pp. 3953–3963, 2010.

[DUP 10] DUPONT J., SCHOLTEN J.D., "On the structural and surface properties of transition-metal nanoparticles in ionic liquids", *Chemical Society Reviews*, vol. 39, pp. 1780–1804, 2010.

[DUP 13] DUPONT J., MENEGHETTI M.R., "On the stabilisation and surface properties of soluble transition-metal nanoparticles in non-functionalised imidazolium-based ionic liquids", *Current Opinion in Colloid and Interface Science*, vol. 18, no. 1, pp. 54–60, 2013.

[DWY 10] DWYER V.M., "An investigation of electromigration induced void nucleation time statistics in short copper interconnects", *Journal of Applied Physics*, vol. 107, no. 10, pp. 103718-1–12, 2010.

[EIN 05] EINATI H., BOGUSH V., SVERDLOV Y. *et al.*, "The effect of tungsten and boron on the Cu barrier and oxidation properties of thin electroless cobalt-tungsten-boron films", *Microelectronic Engineering*, vol. 82, nos. 3–4, pp. 623–628, 2005.

[END 02] ENDRES F., "Ionic liquids: solvents for the electrodeposition of metals and semiconductors", *ChemPhysChem*, vol. 3, no. 2, pp. 2002.

[FED 08] FEDOROV M.V., KORNYSHEV A.A., "Towards understanding the structure and capacitance of electrical double layer in ionic liquids", *Electrochimica Acta*, vol. 53, no. 23, pp. 6835–6840, 2008.

[FED 14] FEDOROV M.V., KORNYSHEV A.A., "Ionic liquids at electrified interfaces", *Chemical Reviews*, vol. 114, no. 5, pp. 2978–3036, 2014.

[FEY 92] FEYNMAN R., "There's plenty of room at the bottom [data storage]", *Journal of Microelectromechanical Systems*, vol. 1, no. 1, pp. 60–66, 1992.

[FLE 66] FLEISCHMANN M., HARRISON J., "The relative roles of adatoms and of ions in solution near step lines during electrocrystallization reactions", *Electrochimica Acta*, vol. 11, no. 7, pp. 749–757, 1966.

[FRE 55] FRENKEL Y.I., *Kinetic Theory of Liquids*, Dover, New York, 1955.

[FRE 11] FREYSCHLAG C.G., MADIX R.J., "Precious metal magic: catalytic wizardry", *Materials Today*, vol. 14, no. 4, pp. 134–142, 2011.

[GAM 06] GAMBINO J., WYNNE J., GILL J. *et al.*, "Self-aligned metal capping layers for copper interconnects using electroless plating", *Microelectronic Engineering*, vol. 83, no. 11–12, pp. 2059–2067, 2006.

[GOE 10] GOESMANN H., FELDMANN C., "Nanoparticulate functional materials", *Angewandte Chemie International Edition*, vol. 49, no. 8, pp. 1362–1395, 2010.

[GRI 01] GRILL A., PATEL V., "Ultralow-k dielectrics prepared by plasma-enhanced chemical vapor deposition", *Applied Physics Letters*, vol. 79, no. 6, pp. 803–805, 2001.

[GUG 85] GUGGENHEIM E.A., *Thermodynamics – An Advanced Treatment for Chemists and Physicists*, North-Holland, Amsterdam, 1985.

[GUT 09] GUTEL T., SANTINI C.C., PHILIPPOT K. *et al.*, "Organized 3D-alkyl imidazolium ionic liquids could be used to control the size of in situ generated ruthenium nanoparticles?", *Journal of Materials Chemistry*, vol. 19, pp. 3624–3631, 2009.

[HAM 12] HAM Y., KIM D., BAEK K. *et al.*, "Metal/dielectric liner formation by a simple solution process for through silicon via interconnection", *Electrochemical and Solid-State Letters*, vol. 15, no. 5, pp. H145–H147, 2012.

[HAN 02] HANAOKA Y., HINODE K., TAKEDA K. *et al.*, "Increase in electrical resistivity of copper and aluminum fine lines", *Materials Transactions*, vol. 43, no. 7, pp. 1621–1623, 2002.

[HAU 06] HAUMESSER P.H., MAITREJEAN S., ROULE A. *et al.*, "Copper deposition: challenges at 32 nm", *Fabtech*, vol. 29, pp. 108–114, March 2006.

[HAY 15] HAYES R., WARR G.G., ATKIN R., "Structure and nanostructure in ionic liquids", *Chemical Reviews*, vol. 115, no. 13, pp. 6357–6426, 2015.

[HEL 12] HELGADOTTIR I., ARQUILLIÈRE P.P., CAMPBELL P.S. *et al.*, "Novel chemical route to size-controlled Ta(0) and Ru-Ta nanoparticles in ionic liquids", *Symposium BBB – Functional Materials and Ionic Liquids*, MRS Online *Proceedings Library Archive*, vol. 1473, 2012.

[HEL 13] HELGADOTTIR I., ARQUILLIÈRE P.P., BREA P. *et al.*, "Synthesis of bimetallic nanoparticles in ionic liquids: chemical routes vs physical vapor deposition", *Microelectronic Engineering*, vol. 107, pp. 229–232, 2013.

[HEL 14] HELGADOTTIR I., FREYCHET G., ARQUILLIÈRE P. *et al.*, "Ru-core/Cu-shell bimetallic nanoparticles with controlled size formed in one-pot synthesis", *Nanoscale*, vol. 6, no. 24, pp. 14856–14862, 2014.

[HES 07] HESSE R.W., *Jewelrymaking through History: An Encyclopedia*, Greenwood Publishing Group, Westport, 2007.

[HU 06] HU C., CANAPERI D., CHEN S., GIGNAC L. *et al.*, "Electromigration Cu mass flow in Cu interconnections", *Thin Solid Films*, vol. 504, nos. 1–2, pp. 274–278, 2006.

[HYD 03] HYDE M.E., COMPTON R.G., "A review of the analysis of multiple nucleation with diffusion controlled growth", *Journal of Electroanalytical Chemistry*, vol. 549, pp. 1–12, 2003.

[IST 02] ISTRATOV A., WEBER E., "Physics of copper in silicon", *Journal of the Electrochemical Society*, vol. 149, pp. G21–G30, 2002.

[ITR 13] ITRS, "International Technology Roadmap for Semiconductors (ITRS) Reports", available at http://www.itrs.net/links/2013ITRS/ Home2013.htm, 2013.

[JOS 12] JOSELL D., WHEELER D., MOFFAT T., "Modeling extreme bottom-up filling of through silicon vias", *Journal of The Electrochemical Society*, vol. 159, no. 10, pp. D570–D576, 2012.

[KAS 00] KASHCHIEV D., *Nucleation*, Butterworth-Heinemann, Oxford, 2000.

[KAT 10] KATTI G., STUCCHI M., DE MEYER K. *et al.*, "Electrical modeling and characterization of through silicon via for three-dimensional ICs", *IEEE Transactions on Electron Devices*, vol. 57, no. 1, pp. 256–262, 2010.

[KIL 64] KILBY J., Miniaturized electronic circuits, US Patent no. 3,138,743, June 23 1964.

[KIM 07] KIM S., DUQUETTE D.J., "Morphology control of copper growth on TiN and TaN diffusion barriers in seedless copper electrodeposition", *Journal of the Electrochemical Society*, vol. 154, no. 4, pp. D195–D200, 2007.

[LAM 50] LAMER V.K., DINEGAR R.H., "Theory, production and mechanism of formation of monodispersed hydrosols", *Journal of the American Chemical Society*, vol. 72, no. 11, pp. 4847–4854, 1950.

[LAN 03] LANE M., LINIGER E., LLOYD J., "Relationship between interfacial adhesion and electromigration in Cu metallization", *Journal of Applied Physics*, vol. 93, no. 3, pp. 1417–1421, 2003.

[LED 08] LEDUC P., DI CIOCCIO L., CHARLET B. *et al.* et al., "Enabling technologies for 3D chip stacking", *International Symposium on*in *VLSI* Technology, Systems and Applications, pp. 76–78, 2008.

[LE 13] LE TIEC Y. (ed.), *Chemistry in Microelectronics*, ISTE, London and John Wiley & Sons, New York, 2013.

[LIN 02] LINIGER E., GIGNAC L., HU C.-K. *et al.*, "In situ study of void growth kinetics in electroplated Cu lines", *Journal of Applied Physics*, vol. 92, no. 4, pp. 1803–1810, 2002.

[LLO 91] LLOYD J.R., "Electromigration failure", *Journal of Applied Physics*, vol. 69, no. 11, pp. 7601–7604, 1991.

[LOT 62] LOTHE J., POUND G.M., "Reconsiderations of nucleation theory", *The Journal of Chemical Physics*, vol. 36, no. 8, pp. 2080–2085, 1962.

[LUC 15] LUCZAK J., PASZKIEWICZ M., KRUKOWSKA A. *et al.*, "Ionic liquids for nano- and microstructures preparation. Part 1: properties and multifunctional role", *Advances in Colloid and Interface Science*, vol. 230, pp. 13–18, 2015.

[LYD 85] LYDE D.R., *Handbook of Chemistry and Physics*, 85th ed., CRC Press, 1985.

[MAI 04] MAITREJEAN S., JOUSSEAUME V., TROUVE H. *et al.*, "Study of ALD and CVD metallization compatibility with ULK using a post integration porogen removal approach.", *Advanced Metallization Conference (AMC)*, pp. 83–88, 2004.

[MAT 59] MATTSSON E., BOCKRIS J., "Galvanostatic studies of the kinetics of deposition and dissolution in the copper + copper sulphate system", *Transactions of the Faraday Society*, vol. 55, no. 9, pp. 1586–1601, 1959.

[MCD 63] MCDONALD J.E., "Homogeneous nucleation of vapor condensation. II. Kinetic aspects", *American Journal of Physics*, vol. 31, no. 1, pp. 31–41, 1963.

[MIL 74a] MILCHEV A., STOYANOV S., KAISCHEV R., "Atomistic theory of electrolytic nucleation: {II}", *Thin Solid Films*, vol. 22, no. 3, pp. 267–274, 1974.

[MIL 74b] MILCHEV A., STOYANOV S., KAISCHEV R., "Atomistic theory of electrolytic nucleation:{I}", *Thin Solid Films*, vol. 22, no. 3, pp. 255–265, 1974.

[MOF 05] MOFFAT T., WHEELER D., EDELSTEIN M. *et al.*, "Superconformal film growth: mechanism and quantification", *IBM Journal of Reasearch and Development*, vol. 49, no. 1, pp. 19–36, 2005.

[MOO 65] MOORE G.E., "Cramming more components onto integrated circuits", *Electronics*, vol. 38, no. 8, pp. 114–117, 1965.

[MUT 13] MUTAFTSCHIEV B., *The Atomistic Nature of Crystal Growth*, Springer-Verlag, Berlin Heidelberg, 2013.

[NEW 82] NEWMAN R., "Defects in silicon", *Reports on Progress in Physics*, vol. 45, pp. 1163–1210, 1982.

[NIS 07] NISHI Y., DOERING R., *Handbook of Semiconductor Manufacturing Technology*, CRC Press, 2007.

[OHN 85] OHNO I., WAKABAYASHI O., HARUYAMA S., "Anodic oxidation of reductants in electroless plating", *Journal of the Electrochemical Society*, vol. 132, pp. 2323–2330, 1985.

[OLI 08] OLIVIER S., DECORPS T., BERNARD M. *et al.*, "Physical investigation of the impact of electrolessly deposited self-aligned caps on insulation of copper interconnects", *Microelectronic Engineering*, vol. 85, no. 10, pp. 2051–2054, 2008.

[OLI 10] OLIVIER S., DECORPS T., CALVO-MUNOZ M. *et al.*, "Inhomogeneous nucleation and growth of palladium and alloyed cobalt during self-aligned capping of advanced copper interconnects", *Thin Solid Films*, vol. 518, no. 17, pp. 4773–4778, 2010.

[PAC 08] PACHON L.D., ROTHENBERG G., "Transition-metal nanoparticles: synthesis, stability and the leaching issue", *Applied Organometallic Chemistry*, vol. 22, no. 6, pp. 288–299, 2008.

[PAD 07] PADUA A.A.H., GOMES M.F.C., LOPES J.N.A.C., "Molecular solutes in ionic liquids: a structural perspective", *Accounts of Chemical Research*, vol. 40, no. 11, pp. 1087–1096, 2007.

[PAN 04] PANIAGO R., DE SIERVO A., SOARES E. *et al.*, "Pd growth on Cu(111): stress relaxation through surface alloying?", *Surface Science*, vol. 560, nos. 1–3, pp. 27–34, 2004.

[PAR 07] PARK K.Y., KIM H.J., SUH Y.J., "Preparation of tantalum nanopowders through hydrogen reduction of TaCl5 vapor", *Powder Technology*, vol. 172, no. 3, pp. 144–148, 2007.

[PEN 11] PENSADO A., PADUA A., "Solvation and stabilization of metallic nanoparticles in ionic liquids", *Angewandte Chemie – International Edition*, vol. 50, no. 37, pp. 8683–8687, 2011.

[POU 63] POURBAIX M., *Atlas d'équilibres électrochimiques*, Gauthier-Villars & Cie, Paris, 1963.

[RAD 98] RADZIMSKI Z.J., POSADOWSKI W.M., ROSSNAGEL S.M. *et al.*, "Directional copper deposition using DC magnetron self-sputtering", *Journal of Vacuum Science & Technology B*, vol. 16, no. 3, pp. 1102–1106, 1998.

[RAD 04] RADISIC A., OSKAM G., SEARSON P.C., "Influence of oxide thickness on nucleation and growth of copper on tantalum", *Journal of the Electrochemical Society*, vol. 151, no. 6, pp. C369–C374, 2004.

[RED 08] REDEL E., THOMANN R., JANIAK C., "First correlation of nanoparticle size-dependent formation with the ionic liquid anion molecular volume", *Inorganic Chemistry*, vol. 47, no. 1, pp. 14–16, 2008.

[REI 01] REID J., "Copper electrodeposition: principles and recent progress", *Japanese Journal of Applied Physics*, vol. 40, no. 4S, pp. 2650–2657, 2001.

[REN 06] RENAULT O., BROCHIER R., ROULE A. *et al.*, "Work-function imaging of oriented copper grains by photoemission", *Surface and Interface Analysis*, vol. 38, no. 4, pp. 375–377, 2006.

[REN 09] RENARD V., JUBLOT M., GERGAUD P. *et al.*, "Catalyst preparation for CMOS-compatible silicon nanowire synthesis", *Nature Nanotechnology*, vol. 4, no. 10, pp. 654–657, 2009.

[ROD 06] RODUNER E., "Size matters: why nanomaterials are different", *Chemical Society Reviews*, vol. 35, pp. 583–592, 2006.

[RUB 08] RUBIM J.C., TRINDADE F.A., GELESKY M.A. *et al.*, "Surface-enhanced vibrational spectroscopy of tetrafluoroborate 1-n-butyl-3-methylimidazolium (BMIBF4) ionic liquid on silver surfaces", *The Journal of Physical Chemistry C*, vol. 112, no. 49, pp. 19670–19675, 2008.

[SAU 10] SAU T.K., ROGACH A.L., "Nonspherical noble metal nanoparticles: colloid-chemical synthesis and morphology control", *Advanced Materials*, vol. 22, no. 16, pp. 1781–1804, 2010.

[SCH 83] SCHARIFKER B., HILLS G., "Theoretical and experimental studies of multiple nucleation", *Electrochimica Acta*, vol. 28, no. 7, pp. 879–889, 1983.

[SCH 05] SCHMELZER J.W., RÖPKE G., PRIEZZHEV V.B., *Nucleation Theory and Applications*, Wiley-VCH Verlag, Weinheim, 2005.

[SCH 07] SCHREKKER H.S., GELESKY M.A., STRACKE M.P. *et al.*, "Disclosure of the imidazolium cation coordination and stabilization mode in ionic liquid stabilized gold (0) nanoparticles", *Journal of Colloid and Interface Science*, vol. 316, no. 1, pp. 189–195, 2007.

[SHA 03] SHACHAM-DIAMAND Y., INBERG A., SVERDLOV Y. *et al.*, "Electroless processes for micro-and nanoelectronics", *Electrochimica Acta*, vol. 48, no. 20–22, pp. 2987–2996, 2003.

[SHN 92] SHNEIDMAN V.A., WEINBERG M.C., "Induction time in transient nucleation theory", *The Journal of Chemical Physics*, vol. 97, no. 5, pp. 3621–3628, 1992.

[SZI 14] SZILAGYI I., SZABO T., DESERT A. *et al.*, "Particle aggregation mechanisms in ionic liquids", *Physical Chemistry Chemical Physics*, vol. 16, no. 20, pp. 9515–9524, 2014.

[TOK 05] TOKUDA H., HAYAMIZU K., ISHII K. *et al.*, "Physicochemical properties and structures of room temperature ionic liquids. 2. Variation of alkyl chain length in imidazolium cation", *The Journal of Physical Chemistry B*, vol. 109, no. 13, pp. 6103–6110, 2005.

[TRE 93] TREMILLON B., *Electrochimie analytique et réactions en solution*, Masson, Paris, 1993.

[TRI 07] TRIOLO A., RUSSINA O., BLEIF H.-J. *et al.*, "Nanoscale segregation in room temperature ionic liquids", *The Journal of Physical Chemistry B*, vol. 111, no. 18, pp. 4641–4644, 2007.

[VER 05] VEREECKEN P., BINSTEAD R., DELIGIANNI H. *et al.*, "The chemistry of additives in damascene copper plating", *IBM Journal of Research and Development*, vol. 49, no. 1, pp. 3–18, 2005.

[WAN 14] WANG F., RICHARDS V.N., SHIELDS S.P. *et al.*, "Kinetics and mechanisms of aggregative nanocrystal growth", *Chemistry of Materials*, vol. 26, no. 1, pp. 5–21, 2014.

[WEI 97] WEINBERG M.C., BIRNIE D.P., SHNEIDMAN V.A., "Crystallization kinetics and the JMAK equation", *Journal of Non-Crystalline Solids*, vol. 219, pp. 89–99, 1997.

[WES 00] WEST A., "Theory of filling of high-aspect ratio trenches and vias in presence of additives", *Journal of the Electrochemical Society*, vol. 147, no. 1, pp. 227–232, 2000.

[YEA 13] YEAP G., "Smart mobile SoCs driving the semiconductor industry: technology trend, challenges and opportunities", *IEEE International Electron Devices Meeting (IEDM)*, Washington D.C., pp. 1.3.1–1.3.8, 2013.

[ZHO 07] ZHOU J., REID J.D., "Impact of leveler molecular weight and concentration on damascene copper electroplating", *ECS Transactions*, vol. 2, no. 6, pp. 77–92, 2007.

Index

A

additives, 74, 81, 105, 159
  accelerator, 74, 76–78, 81–83, 90, 91, 99, 105
  inhibitor, 99–103
  leveler, 89, 90
  suppressor, 74, 78, 79, 81–83
aggregation, 18, 48, 50, 69, 70
approximation
  capillarity, 19, 31, 42
  quadratic, 35, 37
attachment, 19, 21, 22, 24, 25, 27, 39, 44, 46
  sticking coefficient, 45
autocatalytic reaction, 124

B

barrier layers, 6, 7, 113, 121, 125
bottom-up, 12, 163
bump, 89, 90

C

chemical potential, 14, 15, 23, 29, 61
  electrochemical potential, 61, 63
chronopotentiometry, 67, 76, 82, 103
cluster, 16, 19, 20, 27, 29, 31, 51
  concentration, 20–22, 27, 28, 36, 38, 42, 49

  work of formation, 16, 19, 25, 28, 30, 32–34
coalescence, 18, 48
conformal, 86, 87, 91, 92
copper, 4, 6, 8, 71, 109, 132, 133, 148, 150
critical size, 17, 26, 32, 33, 156, 158
cyclic voltammetry, 75, 76, 114
  hysteresis, 81, 100–102

D

damascene, 5, 6, 73, 83, 87, 94
deposition rate, 66, 73, 81, 84, 127
detachment, 19, 21, 22, 24, 25, 28, 46
dielectric materials, 4, 7, 120
diffusion, 45, 93, 95, 105, 156
displacement reaction, 124
DLVO theory, 69, 152

E

electrical double layer, 69, 152
electroless deposition, 121
electrolyte, 59, 73–75, 82, 96, 99, 108, 110, 124, 126, 144
electromigration, 120, 121, 138, 141
electroplating, 5, 7, 59, 60, 71, 96
encapsulation, 7, 119
equilibrium, 22, 24, 27, 28, 60, 62
equimolar, 31, 174

extreme fill, 102, 106

F, G, I

flux, 20, 21, 28, 38, 49
growth, 18, 43, 45, 46, 156, 159, 162
  rate, 47, 52–54, 89, 95, 103, 105
  integration, 2
  3D, 8, 9, 92
interconnects, 2–4, 8, 9
ionic liquids, 143, 144, 162

L, M

Lamer mechanism, 68, 160
lines, 3–5, 8, 137
metal precursor, 68, 147–149, 162
metallization, 7, 8, 73, 107, 118
metastable, 15, 124
monomers, 14, 16, 18, 19, 27, 31, 51,
  64, 68
  concentration, 45
Moore's law, 2, 8

N

nanoparticles, 8, 10–12, 146, 148, 150,
  151
  stabilisation, 70, 146, 150, 155,
  163
Nernst potential, 62, 63, 75, 110, 112,
  122, 123
nucleation, 16, 18, 19, 31, 66, 156
  heterogeneous, 16, 32–34, 46
  homogeneous, 16, 32–34
  instantaneous, 53, 55, 56, 67, 68,
  156
  progressive, 53–56, 67

rate, 19, 25–28, 38, 39, 42, 51, 54,
  156
  sites, 23, 39, 41, 42, 52, 53, 112,
  125, 130
  stationary, 26, 36, 39
  time lag, 39, 41, 53, 54, 66, 127
  work of nucleation, 17, 32, 33, 158
nucleus, 17, 18, 25, 26, 158
  concentration, 26
  nucleus region, 35, 36

O, P, S

overpotential, 64–66, 79, 112, 116
pulse-reverse current waveform, 96, 97
seed layer, 6, 7, 108, 110, 111
selective deposition, 121, 123, 125,
  134, 135
subconformal, 6, 72, 84, 98
subnucleus, 17, 38
superconformal, 6, 84–87, 91, 96, 98,
  105
  CEAC model, 89–92
  levelling mechanism, 89, 96
supernucleus, 17, 25, 38, 48
supersaturation, 15, 17, 24, 63–65, 68,
  116, 158
surface specific energy, 30, 31, 33, 43

T, V, Z

tantalum, 149–151
thin films, 6, 12, 107, 142
through silicon vias, 8, 93–95, 104
top-down, 12
vias, 3–5
viscosity, 93, 151, 152, 155
Zeldovich factor, 38

Printed in the United States
By Bookmasters